# Managing Innovation

# Managing Innovation

## Mining for Nuggets

*John C. Huber*

Authors Choice Press
San Jose New York Lincoln Shanghai

**Managing Innovation**
**Mining for Nuggets**



Authors Choice Press
an imprint of iUniverse.com, Inc.

For information address:
iUniverse.com, Inc.
5220 S 16th, Ste. 200
Lincoln, NE 68512
www.iuniverse.com

ISBN: 0-595-20283-7

Printed in the United States of America

*To Sonia:*
*my wife,*
*partner,*
*and*
*best friend.*

# *Contents*

# *Preface*

What is new in this book? One, we describe better ways to do innovation, especially finding good ideas and selecting the best. That is, managing innovation is mostly about managing ideas. Two, this book contains the first systematic studies of inventors. We show that inventors are rare and prolific inventors are very rare, at about 3% of an R&D laboratory. Three, we describe a method to find prolific inventors more effectively, called the Inventor Profile. Four, we show that there are better ways to manage innovative projects. Five, we offer specific recommendations to improve managing innovation. Put another way, even the best books on managing innovation focus on managing the project itself and don't focus on the important work of discovering good projects, which is the focus of this book.

Who should read this book? Our primary audience is inventors, aspiring inventors, and first-level managers in product development laboratories. These are the people who face the day-to-day decisions of managing innovation. In addition, others will benefit from reading the book and gaining a greater understanding of the challenges facing innovation management. These people include marketing managers, business unit managers, CEOs, and academic researchers.

How am I qualified to write this book? This book is partly based upon my 30 years experience in new product development. My experience includes being an inventor at the bench, laboratory manager, marketing manager, sales manager, and business unit manager. My experience was at 3M Company, widely regarded as a leader in innovation.

I have had a major role in 29 innovative new product introductions, with fewer than 20% failures, substantially better than the average rate of 50%.[1] In addition, I have read over 700 books and over 1,100 articles related to innovation.

Why did I write this book? Although I worked hard at being a good manager of innovation, I was not entirely satisfied with my results. So, I did the first systematic study of inventor's skills and methods. The powerful results of this Inventor Profile are included in this book, along with how to use them. In the past five years, I have published more articles on measuring inventive productivity than any other researcher, and perhaps more than all others combined. Thus, this book has the combination of practical experience, a critical selection of relevant research, and my original research.

The theme of the book is "mining for nuggets." Each topic focuses on the new "nuggets." We will focus on those topics that are not well covered in existing books and articles on innovation. For those readers who wish to dig deeper, there is a list of the most important books in the field, along with some comments about their strengths.

It is always important to say what a book is not. This book does not teach you how to invent. What little is known about teaching creative skills is in Chapter 7. But, this book will help you to find better inventive opportunities and better ways to turn them into successes. This book is not about management as a profession. It is not a book about the broad topics of managing a product development laboratory. This book focuses on managing ideas.

Also, this book is not an academic treatise. Managing innovation is not a logical, linear process; it is fundamentally a creative, dynamic process. This book is not intended for readers who are expecting a scholarly, linear, logical, impersonal development of provable topics. Thus, this book must contain some advice that cannot be proved, but rather is based on experience. This approach also means the tone of the book is direct and personal. My choice of references also reflects this approach. That is, the references cited here are the most readable, recent and comprehensive, in my opinion. I believe that busy managers of innovation don't have the time or inclination to read lots of books and articles. I'm trying to help them by doing the mining for these nuggets, as well.

You'll find another service for busy readers in this book. The references and suggested readings in the endnotes contain all the information you need. Most other books make the reader go through two or three steps to get the complete reference. That is, first you find the author and date. Then you look that up in a reference list to get the book or journal title and other information. This process is cheap to print, but hard on the reader.

Among the nuggets in this book are some famous and not-so-famous quotations that I've collected. In my opinion, they capture the essence of some important ideas. When these are from others, I've included the original author. When I don't know the author, there is no author reference. When these are mine, my initials "jch" are shown. If I have not given proper credit, it is unintentional and I apologize.

# *Acknowledgements*

First and foremost among those who should be recognized are my many co-workers on new product development projects. Their hard work and dedication to excellence gave us many successes from which to learn. And we learned from failure, too. To avoid the embarrassment of neglecting some, I will not attempt to list them all. They know who they are.

Good bosses are everyone's dream. I had some of the best. Those who deserve special thanks are Bruce Torp, Roger Lacey, Wolf Strehlow, Jim Johnson and Tait Elder.

Writing is not easy for me. Sometimes my prose doesn't make my ideas clear. I am especially grateful for those who read the draft of this book and made many helpful suggestions: Tait Elder, Les Krogh, and Marvin Johnson.

This book is the result of a lifetime of experience and learning. Throughout most of my life I have been very fortunate to have the encouragement and wise counsel of my wife, partner and best friend, Sonia.

# 1

# *Introduction: Managing Innovation*

First, let's be clear about what I mean by innovation. This book is about a special kind of innovation: new products. We will not be concerned with other kinds of innovation, such as product positioning or brand management, which concern marketing managers. Nor are we concerned with re-engineering organizations, new corporate structures, or other things that concern CEOs. We are concerned with the work of inventors, aspiring inventors, and the first-level manager in a product development laboratory.

Second, let's distinguish innovation from some related terms. An invention is something new. To be a patented invention, it must be new, useful and unobvious. That is, to be patented, an invention cannot arise from applying common knowledge. Innovation is the process of bringing an invention to the marketplace. Commercialization is the last stage in the innovation process, including sales training, customer training, and merchandising. Product development includes innovation, but also includes non-inventive incremental product improvements.

## 1.1 The Range of Innovations

Figure 1.1 The Range of Innovation for a Notebook PC Manufacturer

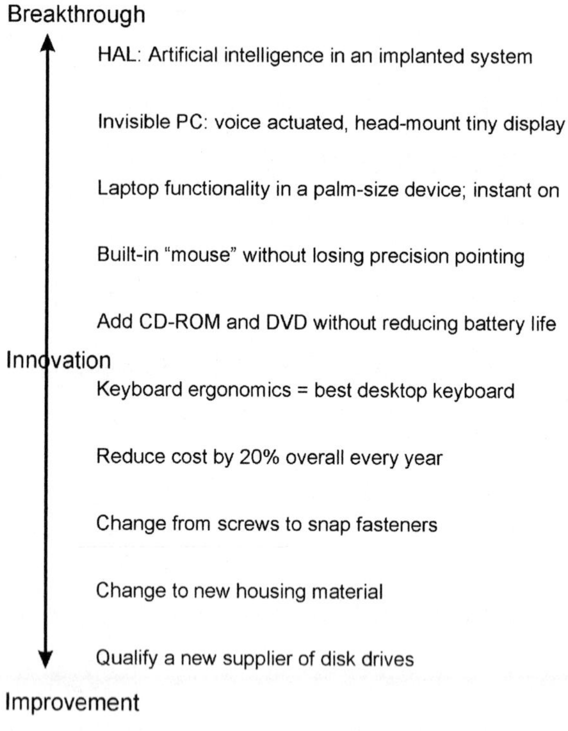

Breakthrough

HAL: Artificial intelligence in an implanted system

Invisible PC: voice actuated, head-mount tiny display

Laptop functionality in a palm-size device; instant on

Built-in "mouse" without losing precision pointing

Add CD-ROM and DVD without reducing battery life

Innovation

Keyboard ergonomics = best desktop keyboard

Reduce cost by 20% overall every year

Change from screws to snap fasteners

Change to new housing material

Qualify a new supplier of disk drives

Improvement

Third, let's be clear about what I mean by new product innovations. In nearly every company, its new products cover a broad range of innovativeness, as shown in Figure 1.1.[2] At one end of the range are incremental product improvements. These are the new products that might be just a change in size, material, or a simple added feature. This work tends to be linear, logical, and relatively easy to manage. Incremental improvements certainly have risks, but the probabilities of their outcomes are known, much like gambling at roulette. Many books describe how to manage incremental product improvements. A list of books appears in Chapter 8 along with my opinion of their strengths.

At the other end of the range are radical breakthroughs. These products are utterly unique and fundamentally transform their markets: examples are the first microprocessor, the first PC spreadsheet. There are very few books that describe how to manage radical breakthroughs,[3] and this is not one of them.

We will be working in the middle part of this range; we'll call it innovation. These products will have some unknowns, but not many. This work has uncertainty. That is, the issues are known, but the probabilities of the outcomes are unknown, much like gambling at poker. Even so, these unknowns are the key to success or failure. Put another way, if there are no unknowns, the problem is incremental product improvement and not innovation. We agree with the dominant opinion that managing innovation is substantially different from managing incremental improvements.[4] That is, risk is different from uncertainty.[5]

Fourth, let's be clear about what I mean by managing innovation. Managing innovation is not just managing people. It is managing ideas, as well. Inventors and aspiring inventors also manage innovation. They manage their own ways of getting ideas and then manage how those ideas become successful new products. However, the first-level manager in a product development laboratory manages innovation by managing people. Innovation cannot be managed in the classical way of command-and-control. Success in innovation cannot be assured by simply employing known rules and principles. So management techniques that are successful in manufacturing or accounting are wholly inadequate in innovation. The uncertainties mean that there will be surprises that cannot be anticipated.

Most importantly, people are not interchangeable in innovation. Therefore, managing innovation is more like coaching a professional sports team. The manager-coach's job is to create a game plan (or strategy) that puts his or her players at a competitive advantage over the opposing team's players. During the game, this strategy may change as the opposing team's game plan becomes clear.

"This game is not about schemes. It is about people. And you have to put your people in the position where they can do their best." John Madden, football commentator.

* * *

Figure 1.2 Components of Innovation

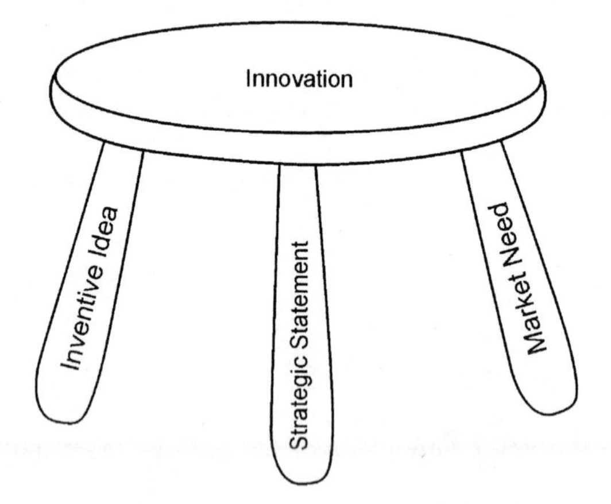


Successfully managing innovation begins when an inventive solution solves an important market problem within a business unit's strategy. The solution must be inventive, otherwise competing companies will quickly copy it and the competitive advantage will be lost. The market problem must be important, otherwise customers won't pay the price needed to cover the costs of developing the product. It must fit within the business unit's strategy, otherwise it will lack the necessary technology, marketing channels, or sales skills. These three components—inventive idea, important market problem, and fit with business strategy—are like legs on a stool; each is important and depends on one another, as shown in Figure 1.2. This book describes how to get this done.

## 1.2 Common versus Uncommon Knowledge

Everything you learned in school, everything written in books, magazines, and everything in training programs is Common Knowledge. Anyone can get it, and it contributes little to competitive advantage. This means that if the technologies, market knowledge, or customer relations of your company can be found in public materials, the company has no competitive advantage. With no competitive advantage, your company is like the prospector panning for gold dust alongside scores of others. Sooner or later, some mishap will knock him out of the business.

* * *

> The only way to guarantee fish for dinner is to buy them at the market. But then nothing is learned about the ways of fish and men. Like fishing for fish, fishing for Uncommon Knowledge is a skill that improves with study and practice.
> jch

* * *

To be successful in managing innovation, you must create Uncommon Knowledge.[6] No one else has it and it can create competitive advantage. A little Uncommon Knowledge can go a long way. SmithKline & French Labs' patent on Tagamet™ gave them an early monopoly on ulcer treatment. Dell Computer's and Wal-Mart's just-in-time manufacturing and distribution give them a competitive advantage in cost and speed. IBM's great strength is not in technology or distribution, but customer relations.

* * *

> "Efficiency is concerned with doing things right. Effectiveness is doing the right things." Peter Drucker, author of many books on management.

* * *

Sometimes Uncommon Knowledge can be clearly described, observed or reverse-engineered. Examples of this kind of Uncommon Knowledge are a company's product designs and market strategies. This is called *explicit knowledge*. However, companies have a vast gold mine of experience, insightful observations, personal relationships, and cultural expectations that often cannot be clearly described. This is called *tacit knowledge*. Entire books have been written about tacit knowledge.[7] Examples of a company's tacit Uncommon Knowledge are hiring and promotion criteria, project selection methods, relations with insightful customers, mutual trust and respect among functions, products that are difficult to reverse-engineer, and products that are patented (though they become Common Knowledge when the patent expires).

* * *

> "The ability to learn faster than competitors may be the only sustainable competitive advantage." Arie De Geus, vice president of R&D at Royal Dutch Shell.

* * *

Of course, the distinction between tacit and explicit knowledge is not black and white. For example, there is a body of knowledge commonly referred to as "know-how." It often begins with the tacit knowledge embodied in a patent. But, as the product and process are further developed, more and more of the knowledge is embedded in the skills and experience of the design and production workers. When this knowledge is passed from generation to generation of workers, it becomes explicit knowledge. When it is very difficult to duplicate, it becomes "know-how" and can be protected as a trade secret. For example, even though the patents for Post-It Notes™ have expired, other companies have not been able to produce competing products.[8]

* * *

> "We can know more than we can tell." Michael Polanyi, eminent scientist.

* * *

The art of finding nuggets certainly requires a firm foundation of Common Knowledge, upon which can be built the fortress of Uncommon Knowledge, as shown in Figure 1.3. While Common Knowledge doesn't provide a competitive advantage, it is also important to maintain its strength. The body of Common Knowledge increases all the time. A company that fails to stay abreast of Common Knowledge risks sinking into the Swamp of Ignorance and Inefficiency. For example, failure to use new management tools, such as Just-in-Time supplier management, Quality Function Deployment, or Total Quality Management may completely negate a competitive advantage gained from Uncommon Knowledge.

Figure 1.3 Fortress of Uncommon Knowledge

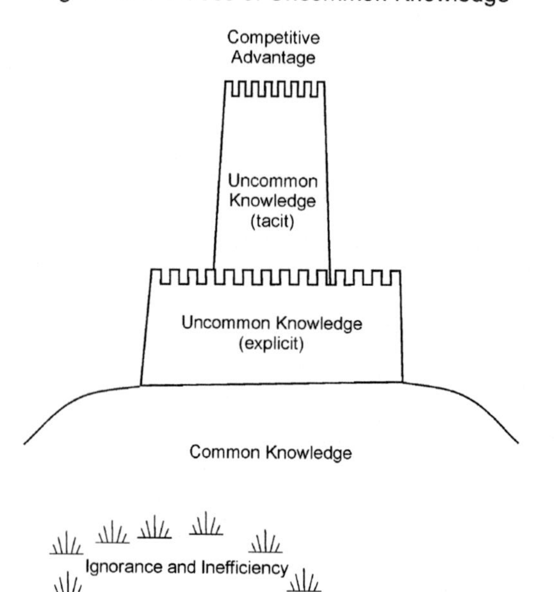


If your Uncommon Knowledge is explicit, then competitors can easily discover it. Then patents are necessary for you to maintain a competitive advantage. If your Uncommon Knowledge is tacit, it is harder for a competitor to discover it, but it is also harder to teach to others in the company. This book will describe many ways for you to discover Uncommon Knowledge and how to leverage it for success.

* * *

> "You have to guess a mathematical theorem before you prove it; you have to guess the idea of the proof before you carry through the details." George Polya, eminent mathematician.

* * *

It is important to be rigorous about what you think is Uncommon Knowledge. For example, the top three competitors in most markets have a combined 90% market share. Whatever knowledge they have in common is not Uncommon Knowledge. That is, it doesn't matter what the weakest competitor doesn't know; it only matters what the strongest competitors don't know.

Throughout this book, we will emphasize the importance of increasing the company's competitive advantage through increasing its Uncommon Knowledge, rather than by merely increasing its R&D expenditures. The overall R&D budget funds many tasks that don't increase Uncommon Knowledge and so have little impact on competitive advantage. Although there are many things that contribute to Uncommon Knowledge, we will mostly describe patents, because they are well-understood, are public documents and have been shown to be indicators of a company's strength.[9]

## 1.3 Summary: Mining for Nuggets

What do I mean by mining for nuggets? I mean that managing innovation is a lot like mining for gold. Gold won't come to you; you must go find it. The easiest gold to find is the gold dust in big rivers. But gold dust doesn't make good pay-dirt. The nuggets get bigger as you go upstream, until finally you find the Mother Lode, where the big payoff lies. All this is hard work; there is no treasure map.

* * *

> Most of the world's water contains no fish. If you want to catch fish, you have to find where they are. Most of the time, the fish aren't feeding. You have to be where they are at the right time. Most fish dislike most foods. You have to

present a desired food at the right time and place. It is the same with innovation. jch

* * *

If managing innovation were easy, there wouldn't be any need for this book or the others on managing innovation. Managing innovation is hard work. One reason is that the components of success described in the preceding paragraph are hard to find. Inventive solutions are hard to find. Important market problems are hard to find. Good strategies are hard to find. Good new product development projects are hard to find. Making the project into a new product success is also hard. Finally, improving future innovations and improving your ways of managing innovation is hard, too.

You wouldn't be reading this far if you were frightened by hard work and unknown problems. The reward is a life lived abundantly. Those poor individuals who seek easy work on known problems live a life of dismal gray. They die before they have lived.

This book will guide you to making the best of your strengths and opportunities. It is a "how-to" book for mining nuggets. This book will help you find where your nuggets are or how to tell them from simple gold dust. However, that is different for every innovation. I call it Uncommon Knowledge.

The next three chapters appear in sequence because that is how books are printed. In real life there is a dynamic interchange among inventive ideas, important market problems, and good strategic statements.[10] You must accept this messy flow. It may take as many as a dozen exchanges to reach a good description of the proposed project. Furthermore, there may be more changes before the innovative new product reaches customers' hands.

# 2

# *Mining for Important Market Problems*

In this chapter, we will show that important market problems are rare, just as gold nuggets are rare. That is, the problems that are easy to find are the same ones your competitors know about. You want the problems that are hard to find, but are so much more valuable.

Let's be clear about what I mean by important market problems. I do not mean simple product improvements; these are not innovations, as defined in the previous chapter. Important market problems are ones that existing products do not solve or they solve them quite poorly. However, we also exclude market problems that would create an entirely new industry if they were solved. The reason is that it takes too much time and money to create new industries.

We are focusing on projects that can contribute to a company's success in less than three years. For example, a company or business unit with $100MM yearly sales would be seeking market problems that could contribute around $5MM yearly in about three years. Obviously, there is a great range in these values. However, a factor of 10 less would be too small to have impact. Similarly, a factor of 10 greater would take too long and cost too much to develop. Don't get bogged down in a sales forecast at this early date. As we will discuss in a later chapter, most new product forecasts are wrong, very wrong or entirely wrong. However, there is a good solution available when a sales forecast is truly needed, as discussed in Chapter 6.

## 2.1 Finding Market Problems

A useful test for identifying an important market problem is that people sit up in bed in the middle of the night and say, "Why can't I get a solution to this?" An example is 3M's Post-It Notes. Previously, people wrote notes on slips of paper and attached them to a document. However, these tended to tear off and get lost. That solution was cheap, but ineffective. By contrast, the Post-It Note was cheap and effective. Another example is the personal computer spreadsheet. The previous solution was either the cumbersome pencil and calculator or a share of an expensive and inconvenient big computer.

* * *

> "Seeing what a customer needs before he knows it himself is the essence of innovation." Art Frey, inventor of Post-It Notes, speaking at the first 3M Technology Update, July 11, 1990.

* * *

Important market problems won't be found in a competitor's catalog. While you are developing your copy of their existing product, that competitor is developing its replacement. Copying tends to put you a generation behind the leader. In the world of industrial products, customers often solicit new products by Request for Proposal or Request for Quotation. Since these documents are usually given to all the possible suppliers, they contain only Common Knowledge. That is, all the suppliers have the same information. A company has a competitive advantage then only if its inventive solution is based on Uncommon Knowledge.

* * *

When you follow the herd, you walk in manure. jch

* * *

You might expect that analyzing market conditions might uncover important market problems. This is The Flaw of Analysis. Analysis is a study of what was. That situation took time to clarify. The data took time to collect and organize. During that time the situation changed. Analysis is a good tool only for competing with the past.

Traditional market research has been effective for identifying and defining incremental product improvements. You also might expect that traditional market research can identify important market problems. Unfortunately, there is little evidence that it is effective. It appears that most customers can offer little guidance about their needs beyond simple changes.[11] Notice that this statement says "most customers." There are a very few customers who are insightful enough to describe important market problems. The first problem is finding these visionaries. The second problem is sorting out the practical visionaries from the dreamers.

In my experience, only about 10% of customers can give you an important market problem when you ask them what are their most important problems. Half of them cannot give you anything but bigger, smaller, faster, cheaper, or more features. These may be useful incremental product improvements, but they are not going to lead to innovations. Then about another half will describe problems that are impossible, very difficult, far out, or a bad match to your strategic statement. Figure 2.1 shows these relationships graphically. It is hard to find the one customer with an important market problem through the noise of nine others.

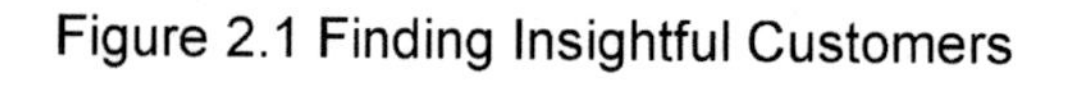
Figure 2.1 Finding Insightful Customers

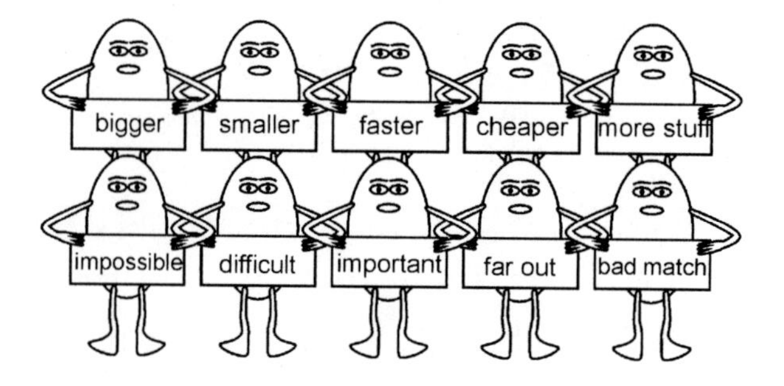


## 2.2 Finding Insightful Customers

The best window to these insightful customers is a courageous marketer. Let's be clear about what I mean by marketer. A marketer is the person whose job is to find new products or new customers. In contrast, a salesperson's job is to sell more of today's products to today's customers. There are administrative tasks in marketing that include keeping track of how much of what product is sold to whom. But our focus is on this courageous marketer. It is important to emphasize how much courage is needed to discover important market problems. First, the marketer must sort out a group of insightful customers. Unfortunately, they are probably as rare as prolific inventors, perhaps less than 10% of all customers. Then the marketer must talk to them.

The sales representative assigned to an insightful customer is likely to be nervous about this contact. Because, if the customer does offer a problem, there is an implied promise to do something about it. As we shall see in the next chapter, most new product development projects are terminated even before a prototype is made. So, the odds of any customer's suggestion being solved are quite small. Even if the marketer does discover several important market needs, it takes courage to propose them as potential new product ideas. Since these problems may not have effective solutions, there will be many questions that don't have answers. How many people have the same problem? Exactly what kind of solution will they accept? How much will they pay? Why hasn't some competitor solved it already? Does the person who pays for it also get the benefit?

This last question is often overlooked. There are lots of problems in the public sector where a worker gets the benefit but the taxpayer pays for it. It often takes a long time to persuade both parties. An example is breathing apparatus for fire fighters. The death and disability problem was known for centuries. This scuba-like product was technically possible for decades, but it was decades longer before budgets were increased enough to pay for them.

* * *

> "Why is marketing so uncreative? Because they cannot do two incompatible things at once. They must listen to the customer about the existing products. The changes needed to keep the short term revenues flowing are vital to the enterprise. This leads to the false assumption that listening to the same customer will give insight into innovative new products. It never does." William Davidow, the marketer that positioned Intel to dominate the microprocessor market.[12]

* * *

An effective way to discover important market needs is to live the life of the customer. Watch what they do. Pay attention to their frustrations. Be like a good physician; ask where it hurts. If you can't do this personally, talk to the people who do. Talk to those people who train customers to use your existing products. Trainers know which parts of the course are hardest to teach and which products are hardest to master. Sometimes sales representatives can be a good source.

What do you do if this approach doesn't produce enough important market problems? One approach is to analyze existing products for fatal flaws. This is the Weak Link Law. If you are a student of history, you know that the French lost Canada to the British because they overlooked a goat track up a supposedly unclimbable cliff. It makes no difference how big the obstacle is. It is only as strong as its weakest point. It can be reduced completely by finding the goat track.

An example may be helpful. I worked in a business unit that sold electrical connectors for splicing telephone wiring. When the copper wiring began to be replaced with optical fibers, we knew we needed to have an optical fiber splice. We bought some of the most common optical fiber splices and searched for their weaknesses. The most popular splice was essentially a long pencil eraser with a small hole down the axis. The two fibers to be spliced were pushed down the opposite ends of the hole until they met in the middle. The hole had to be smaller than the fibers to hold them firmly joined, even during vibration and changing temperature. However, this meant that the hair-thin glass fibers frequently broke when being pushed into the hole.

We were indeed fortunate to have prolific inventors volunteer to work on this problem. Jack Blomgren's inventive idea was to make the hole open and close. When open, the hole was slightly larger than the fibers and they slipped in easily. When closed, the fibers were gripped even more firmly. Fiber breakage was eliminated and other performance measures improved. Dick Patterson's inventive idea was to use a better mechanism to open and close the hole and better materials for a stronger grip on the fiber. The combination of these ideas gave dramatically improved performance. Our patented product dominated the market in less than a year, and sales of the competitor's splice declined by 30% per month. Our splice is still the dominant product of its type even after fourteen years. This is the power of Uncommon Knowledge. You can develop it by using the Weak Link Law.

Finally, don't stop at the first good problem. Most important market problems don't have effective solutions. Most problems with solutions won't fit the company's strategy. Most of those that pass these tests won't have a good chance of meeting all the requirements of a good project, as discussed in a following chapter. Most new products fail in the marketplace. So you need a lot of good problems to have one survive to make a contribution to the company.

* * *

> "Great marketing requires total commitment. You have to believe. You have to create. You have to sacrifice. You have to do battle. You have to win. Great marketing—I call it Total Marketing—is a crusade. If you aren't strong enough to stay in the field, if you can't provide your product with a soul, you should find a nicer game to play." William H. Davidow, Intel marketer who established their microprocessor as the industry leader.[13]

* * *

## 2.3 Evaluating Important Market Needs

Figure 2.2 Head-Slap versus Foot-Shuffle

How do you know which problems are the best? This brings us to the head-slap or foot-shuffle law of innovation, shown in Figure 2.2. When you describe the problem to someone else, do they slap their heads and say "Wow!" or do they shuffle their feet and say "Hmm that's interesting." If you get mostly foot shuffles, it's probably not an important problem. Please notice that this law is being applied to describing the problem, not the solution to the problem. That is a different situation that we will discuss in a later chapter.

If the responses are mostly "Wow!" it is a lot easier to get stakeholders. The first and most important stakeholders are other customers. Even a moderately insightful customer can respond to the description of a problem. If a marketer can say that lots of customers said they had the same problem, it does wonders for his or her courage. Another important group of stakeholders are the prolific inventors. If they say "Wow!" it is a lot easier to get them to offer their inventive ideas for a solution. If both customers and inventors say "Wow!"

upper management is a lot more comfortable about making a small investment. If all three groups say "Wow!" it's a lot easier to get help from the sales force to check the problem with more customers.

However, it is important to keep a low profile in the marketplace during this period. All your competitors are looking for important market problems, too. Once the problem is clearly described, many customers will share it with other suppliers, just to get the benefit of more competition. This is almost an article of faith among purchasing agents.

One way to maintain secrecy is to have a non-disclosure agreement with the insightful customers. However, this approach has two drawbacks. First, the terms of the agreement may impact your ability to get a patent. Don't get bogged down in this complex area, consult your legal department. But don't let them be so legal-bent that they inhibit the business. Remind them that the goal is to create a successful new product; without this feedback, the product will fail, often very expensively. Second, the customer's legal department gets involved and this can create substantial delay. As I said before, this managing innovation is hard work!

## 2.4 Summary

The nugget of knowledge in this chapter is to be selective about deciding who are the insightful customers. Listen to them very carefully. Watch for head-slap opportunities.

# 3

# *Mining for Inventive Ideas*

* * *

"All imaginative inventions are errors from the norm." Jacob Bronowski, mathematician, scientist, and author of the best-selling book *The Origins of Knowledge and Imagination.*

* * *

Remember our three-legged stool of managing innovation? One of the legs is an inventive solution. If you have read other books on managing innovation, you might have noticed that not much is said about where good ideas come from. Why? Inventive talent is a very rare, but vitally important type of Uncommon Knowledge. Inventive talent is a rare nugget, indeed.

What do I mean by an inventive idea? The dictionary defines an invention as something new. However, it is more valuable if the invention is new, useful and unobvious, and thus has the potential to be protected by a patent. Without going into the complexities of patent law, a U.S. patent gives the inventor a 20-year monopoly on the invention.

## 3.1 The Rarity of Inventors

How rare are inventors? Of the adult population, less than 1% have at least one patent.[14] Of research and development workers, less than 36% have at least one patent.[15] Of all inventors, only about 10% have at least five patents over at least four years; we call these the *prolific inventors.*[16] However, this 10% of inventors are responsible for about half of all patents![17] In other words, about 0.1% of the people produce half the patents! Thus, inventors are rare, and prolific inventors are very rare indeed.

Figure 3.1 Lotka's Law

percent of inventors versus number of patents

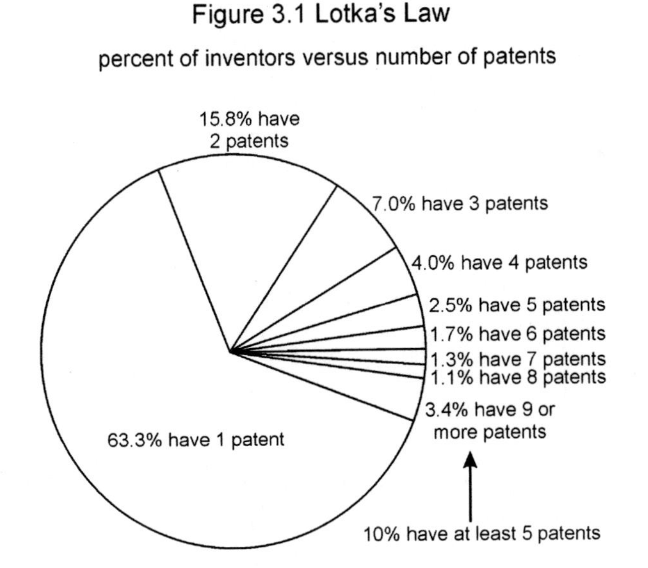


To get another picture of the rarity of inventors, see Figure 3.1. This chart shows the percentage of inventors versus increasing numbers of patents. It shows that only 10% of inventors have at least five patents.

This rapid fall off is also typical of scientific publications by physicists, and chemists, and is called Lotka's Law.[18] The rarity of inventors goes as the -1.4 power of their number of patents. Thus, only about 10% of inventors have at least five patents. However, the total number of patents held by an inventor mixes together the rate of production with the length of the inventor's career. When we examine the rate of patent production, we find that most inventors have a low rate of patent production and very few have a large rate, as shown in Figure 3.2. The rarity of an inventor's productivity goes as 2.72 to the negative power of that inventor's patent productivity divided by 0.7.[19] Thus, only about 6% of inventors produce over two patents per year. What is surprising is that these relationships don't vary much among companies.[20] This indicates that companies don't differ much in the ways that they hire and manage inventors. This book offers some ways to improve this situation.

Figure 3.2 Patent Productivity

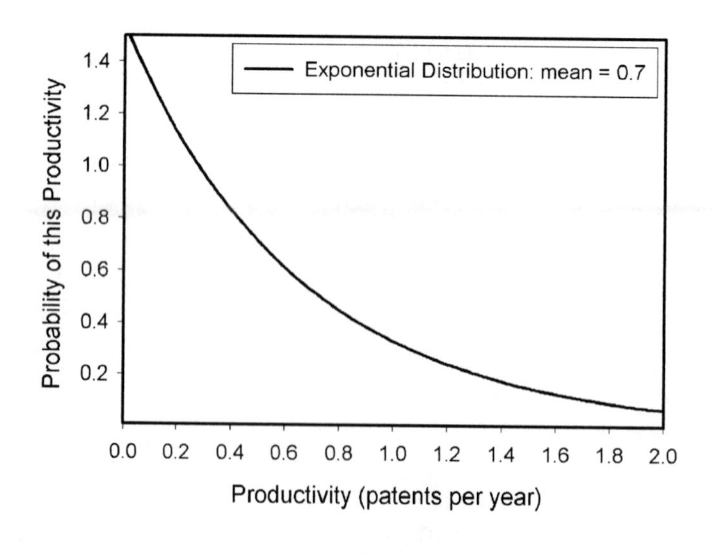


* * *

> “The scarcest resources in any organization are high-performing people.” Peter F. Drucker, author of many best-selling books on management.

* * *

An important issue is patent quality versus quantity. Of course, patent quality is very difficult to evaluate from public information. But the cost of filing and maintaining a patent is substantial and mitigates against producing low-value patents. The average cost of filing a U. S. patent is $ 7,000.[21] However, multinational companies need to protect their inventions around the world; 3M reports an average cost of $140,000 per patent, not including any litigation.[22]

We can put this in concrete terms. Suppose you are the lab manager for a company or business unit with $100MM in yearly sales, with 5% invested in R&D. That is a $5MM yearly budget. However, the typical lab cost per person is about $140,000 per year: including salary, benefits, equipment, space, and administrative support. So this lab manager has about 36 people who might be inventors. Based on the averages described above, only 13 of them will have at least one patent. Only one or two people will have five or more patents, but they will produce half the patents.

* * *

> “It is in the mind of a single person that creative ideas and concepts are born.” Marvin Kelly, Director of Research at Bell Labs during the development of the transistor.

* * *

## 3.2 Finding Inventive Ideas by Inventors

If you want to get the competitive advantage of a patent, your invention must be new, useful and unobvious. The first two criteria are usually straightforward; it is the last one that is the hardest. Your technical training taught you to build your research on the Common Knowledge of others. The result of this kind of research tends to be incremental product improvements. If you want innovations, you have to look at the important market need and the available technologies differently from the other people. You have to see what they don't see. You have to imagine what they cannot. The preceding section on the rarity of inventors shows that these capabilities are rare. How can this be improved?

There are lots and lots of stories in science and invention about the powerful imaginations of the great geniuses. Einstein imagined himself riding on a beam of light. Tesla imagined building electrical equipment in his mind and then examining it for vibration and wear. Kettering imagined what it was like to be a piston in an engine. There are also lots and lots of stories about sudden inspirations, ideas that suddenly pop into the mind. The hexagonal structure of benzene came in a dream. Are these stories true? Just how important are imagination and inspiration for invention? Does it matter? And how much?

The only systematic study of this question is the Inventor Profile, described in Appendix B. Inventors who imagined they were the product had a larger rate of patent production. The relationship is very strong; the odds are 75:1 against it being a coincidence. The inventors who did this imagining also had more inspirations and sudden insights. The relationship is strong; the odds are 7:1 against it being a coincidence. However, this imagination was still rare; 84% of these inventors never imagined they were the product. And this study examined the top 2.5% of all inventors!

So the conclusion is clear. Imagination matters and it matters a lot. Imagination leads to more inspirations and more patents. But this intensive imagination is rare. The implication is also clear. If you want to have more inventive ideas and better inventive ideas, you need to engage your imagination. There are some books that describe ways to improve your imagination listed in the endnotes; my favorite is Shore's.[23] However, none of these books show how much these techniques actually improve your imagination.

Finding inventive ideas is not limited to your working alone. Prolific inventors are often eager to kick around ideas with other inventors. The synergy often creates remarkable improvements. Perhaps the most successful example is the Post-It Notes story. Put very briefly, Spence Silver invented the not-so-sticky adhesive. Art Fry invented the application. It was Spence Silver's persistence in sharing his idea that created the vital linkage.

## 3.3 Finding Inventive Ideas by Managers

So, if you want to have inventive solutions, they are most likely to come from a very few people. Remember that we are mining for nuggets. The gold won't come to you. You must find where the gold is among these few people. Some of them will be fully absorbed in other critical projects. Some will lack the necessary skills. There are a few good books on managing these rare and sometimes difficult people.[24]

* * *

> The manager is not the most important person; the inventor is. jch

* * *

If you are the kind of person who got into management because you are attracted to personal power, you are in for a rude surprise in managing innovation. As we said in the previous chapter, a successful manager of innovation is like a coach in professional sports. Your job is to create a situation in which your players can win. You must let them be the stars. Your reward is when they are successful.

* * *

> Inventive people do not respect ordinary power. However, they do respect commitment and accomplishment. A manager can earn their respect by providing them resources when they are needed and relieving them of tedious organizational duties. jch

* * *

In the beginning, some of these ideas, needs and strategies will be crudely described and will not fit together smoothly. Moreover, these components do not follow each other in a nice, linear sequence; they interact with each other. A beginning must be made, and one place is not necessarily better than another. This is the most fragile moment in innovation; there are many unknowns. The manager's most important contribution is creating the environment, also known as the *people side of innovation.*[25]

* * *

> The manager's job is to create an environment where inventive ideas meet important market needs that fit the company's strategy. jch

* * *

In my experience, a fruitful place to find ideas is with each inventor's backlog of ideas. As we will discuss below, there are lots more ideas than there are projects. Every inventor has some ideas that he or she thinks are good ones but have not been accepted. Again, in my experience, the best way to discover these is to invite the inventor to lunch and broach the topic directly, such as, "I sure wish we had some new ideas to work on around here. In a few months, the big projects will all be finished. It would be nice if we had a few coming along on the practice squad." (Sports metaphors tend to play well in innovation.) At this fragile stage, always be at least moderately positive about an idea. There is a lot to be learned about these ideas that is not known at this time. Conversely, it is not very productive to send a memo and ask for written responses. As a group, inventors hate to write, especially about something as fragile as a new, untried idea or an idea that has already experienced some rejection. Furthermore, written responses tend to favor those ideas that are more well-developed and look good on paper; these may not be the best ideas.

## 3.4 Finding Prolific Inventors

Obviously, the best way to identify prolific inventors is by their rate of patent production.[26] However, many of the people in a lab may not have had the opportunity to invent. How can these potential inventors be identified? We have created the Inventor Profile to discover these potential inventors. It is discussed in detail in Appendix B. If you want a copy of the survey format, send me an email at jchuber@InventionAndInnovation.org and I will send it to you in electronic form, ready to make into a booklet.

The Inventor Profile is based on published indicators of inventive potential and my 28 years of experience in managing inventors.

Surprisingly, this is the first systematic collection of the voice of the inventor! Prolific inventors' general characteristics are not surprising.

- They have a strong desire for excellence (61% chose their field to "be the best").
- They are well-educated (45% with graduate degrees with honors; 90% have an undergraduate GPA of 3.0 or higher).
- In school, they spent little time away from their studies (average 5.6 hours per week). Hard work is common (average 50 hours per week).
- They are team players (56% preferred working on a team).

However, some of the common beliefs about inventors are found to be either non-predictive or even negative predictors. That is, time spent playing a musical instrument, chess or bridge games, or reading books outside their coursework are predictors of generally lower patent productivity! Perhaps most surprising is that the highest undergraduate GPA is a predictor of generally lower patent productivity, though 90% had a 3.0 or better and 45% had graduate degrees with honors. A possible explanation is that inventors' inquiring minds and eagerness to experiment tend to interfere with their getting the highest undergraduate GPA, but fit well into the more investigative graduate program.[27]

In developing the Inventor Profile, we also discovered that some common beliefs about generating creative ideas are subject to dispute. Less than 12% of these top inventors had a positive view of brainstorming. Less than 3% of them used visualization to enhance their creative ideas. Many books and articles have implied that they should use these techniques, but they do not. One possible explanation is that these inventors are more comfortable with logical problem solving, as opposed to a more spontaneous or intuitive approach.

The Inventor Profile focuses on objective facts and talents. However, there are some subjective characteristics that help to identify prolific inventors:

- a healthy discontent with existing products,
- courage to advance their ideas in the face of discouragement,[28]
- dogged persistence.[29]

Figure 3.3 Predicted versus Actual Patent Productivity

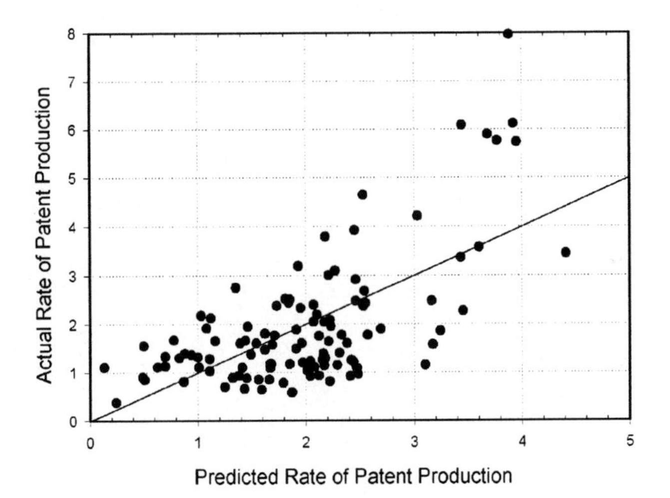


All the variables taken together explain 43% of the variance in the actual rate of patent production, as shown in Figure 3.3. The principal findings from the Profile are summarized in Figure 3.4. The variables with a negative sign (-) are inversely related to patent productivity. That is, those inventors who spent a lot of time practicing a musical instrument had fewer patents per year than those who did not. The Figure shows that just 12 variables are the strongest predictors. Having an advanced degree with honors predicts 25% of an inventor's patent productivity. That, combined with working many hours per week, predicts 37%. Each further variable contributes less and less. Figure 3.5 shows these variables rearranged, with those appropriate for new hires to the left and those for experienced people to the right. About half of an inventor's patent productivity is predictable when they are new graduates and the other half when they have substantial experience.

Figure 3.4 Effect of Variables in Order of Impact

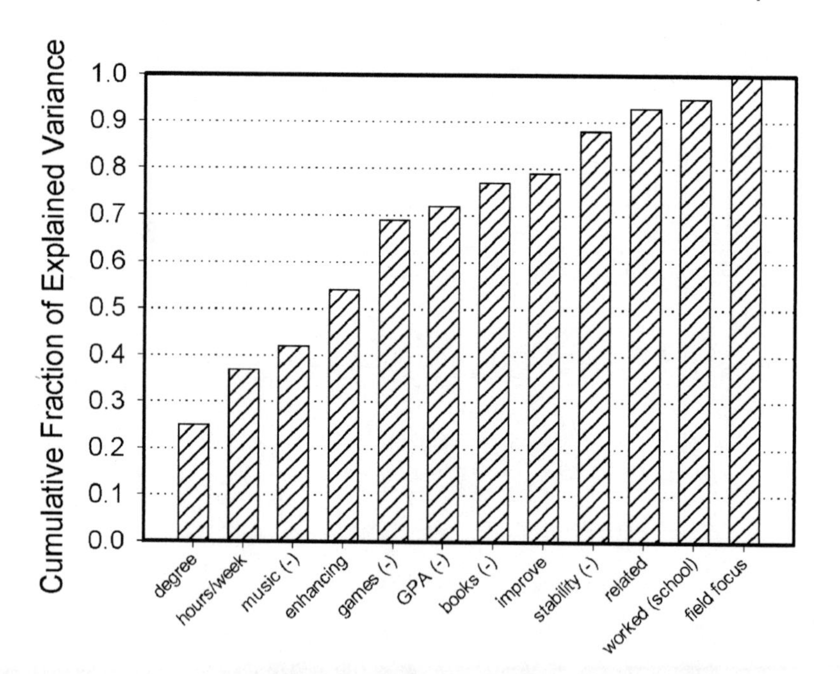


Variables in Descending Order of Impact

Figure 3.5 Effect of Variables in Order of Experience

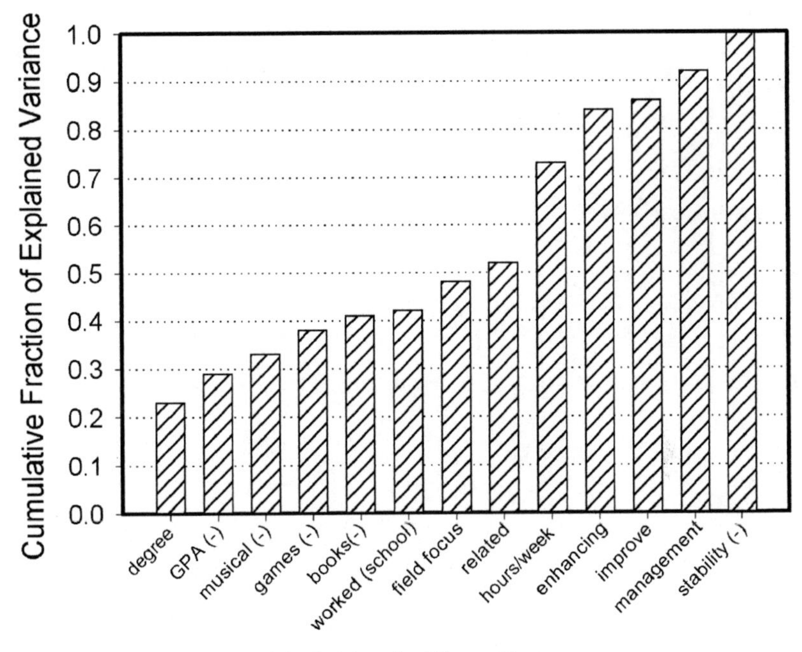


At this point, we are interested only in identifying existing prolific inventors. Give the Inventor Profile to your lab people and ask them to complete it. It should take them only about 15 minutes. Score the profile. Those with the highest scores are the most likely to become your most prolific inventors. For suggestions on what to do with the less-inventive people and hiring replacements, see Chapter 7.

Psychologists have been studying what makes people creative for the past 50 years. They have had little success.[30] Why do people invent? They just do. It is the same reason as to why people dance, draw, sing, or climb mountains. These are the behaviors that fulfill their psyches.

They love it. Most know it from early childhood. Occasionally, someone discovers this love later in life.

The goal is to identify the most prolific inventors in your lab. They are the human nuggets among the ordinary. These human nuggets will generate a wealth of inventive ideas, the nuggets of Uncommon Knowledge.

## 3.5 Summary

The nuggets of knowledge in this chapter are that inventors are rare, but they can be identified with the Inventor Profile. Inventive ideas can be enhanced by improving your imagination. Most inventive ideas will come from inventors who are prolific.

# 4

# *Mining for Good Strategic Statements*

Let's be clear about what I mean by a good strategic statement. Nearly every company has a strategic plan, mission statement, or vision statement. However, the vast majority of them are self-congratulatory platitudes about the existing business. They talk about quality, customer service, and employees as major assets. Very few of them talk about where they want to be in the next five years.

* * *

> "If you don't know where you are going, any path will take you there." Chinese proverb. And you won't know when you have arrived. jch

* * *

For example, a good strategic statement for 3M's Post-It Notes business unit would be, "We will continue to be the dominant supplier of document annotation products." I'm not revealing any secrets here; I have no knowledge if this is their plan or not. But if it isn't, it should be! As another example, a good strategic statement for a notebook PC manufacturer would be, "We will be the dominant supplier of easy-to-use and easy-to-carry computers with power equal to a mid-range desktop computer."

A good strategic statement should be both tight and loose. The Post-It example above would include ways to annotate electronic documents, whether in an individual's personal computer or on the Internet. Yet it excludes ways to create, copy or store documents. It possibly could include ways to prevent documents from being altered, and thus preventing annotation. The notebook PC example would include ways to migrate toward palm-size devices, but the computing power requirement would avoid direct confrontation with palm-size devices.

* * *

It is not enough to plan to survive. You must plan to thrive.
jch

* * *

What if your company or business unit doesn't have a good strategic statement? First, don't blame the CEO. He or she is totally absorbed in this quarter's sales and profits. Some CEOs are so focused they interrogate their subordinates about daily sales! I know; I worked for one. In addition, many CEOs would be reluctant to be too clear about a forward-looking statement. It can get into the hands of competitors and strengthen their positions. If it is necessary to change it, he or she might be criticized by the financial press or stockholders.

However, if you're not part of the solution, then you're part of the problem. Take the issue in hand and write your own strategic statement.

Talk about it with your colleagues. Listen to their comments and suggestions. Here is another place where the head-slap or foot-shuffle law can help. If the strategic statement has mostly "Wow!" responses, you're on the right track.

It is especially important that all the company's functions support the strategic statement—laboratory, marketing, sales, manufacturing. For example, if manufacturing has total focus on improving quality, it will be reluctant to take on a new product that uses a new manufacturing process. It is important to get everybody's goals out in the open at this early stage. You certainly don't want to find incompatible goals after you've invested years of effort.

However, the statement must be compatible with your company. If it calls for dramatically new technologies, marketing channels, manufacturing processes or sales skills, many people will not support it.

Always keep in mind that you want a one-sentence statement, not a complete plan. It is easy for this statement to grow into many pages. Leave the big plan for later.

Finally, do not fail to do this. If your company lacks a good strategic statement, your chance of having a successful new product is nearly zero.[31] Without it, you have no way of choosing good inventive ideas, important market problems, or good projects.

# 5

## *Mining for Good Projects*

When an inventive idea appears to solve an important market problem and fits within the company's strategy, it becomes a proposal. That is, the evolutionary sequence is: inventive idea, proposal, project, new product. If you've gotten to this point, you have several possible proposals. Congratulations! But you are far from having a success. It takes about 3,000 good ideas to generate one successful new product.[32] Even with the fit to the market and strategy, your odds are still less than 100:1. Fortunately, there are usually more good ideas than a company has resources to invest.

* * *

> "To invent is to choose." Jaques Hadamard, eminent mathematician. "To govern is to choose." Winston Churchill, Prime Minister of the United Kingdom during World War II. Obviously, to manage innovation is to choose. jch

* * *

Let's revisit our example company. There are 36 people in the laboratory. Probably half of them are working on product improvements.[33] Probably half of the remaining are working on innovations that are well on their way to becoming successful products. That leaves about nine people, about enough to staff two or three new product development projects. How do you decide which of your proposals actually will become projects?

This period of proposal development is called the Fuzzy Front End.[34] At this point, you don't know what it will take to have a successful product. The situation has many uncertainties. You don't really know if the inventive idea will solve the problem. You don't really know how many customers will buy. You don't really know how much they will pay. You don't know how much it will cost to make the product. Uncertainty is different from risk. Risk is when you know the question and the probabilities of its outcomes, but the outcome varies, similar to gambling at roulette. Uncertainty is when you know the question, but not the probabilities of the outcomes, similar to gambling at poker. Moreover, you really can't know these things until you have put a prototype in a customer's hands and received a response. Sometimes, you will find that you are working on the wrong questions, which is a high level of uncertainty. The main purpose of product development is to reduce uncertainty.[35]

## 5.1 Selecting Good Projects

There are probably hundreds of methods for project selection, also called prioritization. Most of them are some variation on net present value and require good estimates for sales and costs. However, at the Fuzzy Front End, we don't have even weak estimates! At this stage, most new product forecasts are wrong, very wrong, or entirely wrong. The old computer saying applies here: Garbage in, garbage out. New product success or failure is largely decided by the first few decisions, those that precede actual product development.[36] Using net present value at this early stage weakens the quality of these decisions. Experienced managers and researchers all agree that managing the Fuzzy Front End is fundamentally different from managing the rest of the new product development process.[37]

* * *

> "The typical practice of discounted cash analysis deals with [innovation] either poorly or not at all. Discounted cash flow, internal rate of return, or net present value techniques—all the same thing with different names—are particularly unable to tell you the consequences of not doing something innovative; the tool is ill suited to the task and its recommendations should be regarded with suspicion by managers who want to succeed at commercialization."
> Donald Frey, CEO of Bell & Howell[38]

* * *

Management's primary function is to invest resources before all the facts are available. This is doubly true for the management of innovation. By the time all the facts are known, they are Common Knowledge. By then some bold and courageous competitor has already created the necessary Uncommon Knowledge and given itself a competitive advantage.[39]

* * *

"It is usually too early before it is too late." Gunther Dierssen, prolific inventor.

* * *

What is needed at the Fuzzy Front End is a method to assess an idea's uncertainties and its contribution to Uncommon Knowledge.[40] The best project selection method I've found for the Fuzzy Front End is Cooper's NewProd™.[41] A major chemical company reported very good success with a slightly modified version. I've used a very similar, but proprietary, method for over 20 years and with very good success.

Figure 5.1 A Project Selection System

| Success Factor | Score |
|---|---|
| How well is the product performance defined? | |
| Performance definition is written and approved | 2 |
| The function is known | 1 |
| Concept only | 0 |
| What is the likely technical advantage? | |
| We will have a strong patent | 6 |
| The technology is difficult for others to copy | 3 |
| The technology is easy to copy | -4 |
| Is the technology similar to other products we have? | |
| Major items are similar | 6 |
| Minor items are similar | 3 |
| Very different | -2 |

These methods ask specific questions about the most important factors for success. Each question has a few specific multiple-choice answers. Each answer produces a numeric score. An example is shown in Figure 5.1. You should test the project selection method on your company's new products, both successes and failures. A sample of the 10 or 20 most recent innovative new products is all that is necessary. You may need to modify the questions, adjust the ranks, or apply weights. Work on it until it gives about 60–80% correct estimates for your successes and failures. This work should be a joint effort, including laboratory, marketing, and manufacturing. It may take several man-weeks to do this work, but it may save several million dollars in new product failures.

* * *

Anna Karenina's Principle of New Products: Successful products are all alike. Unsuccessful products are all different in their fatal flaw. (Therefore, a project selection system should be designed to find these flaws.)

* * *

The best project selection systems assign numerical scores to various questions about the proposal. However, some people's ways of thinking and making decisions make them skeptical of numerical methods. In that case, a list of questions can be used.[42] After having this discussion with the project champion, a group of evaluators can decide whether the proposal will become a project.

* * *

"The best way to predict the future is to create it." Charles F. Kettering, Vice President of Research, General Motors

* * *

It is important to keep in mind that a long, detailed description of a proposal is generally not related to its quality. A brief, clear description is all that is necessary.

* * *

"It's easy to say what you're going to do. The hard part is figuring out what you're not going to do." Michael Dell, Founder and Chairman of Dell Computer Corporation

*   *   *

It takes about 100 inventive ideas and important market problems that match your company's strategic statement to get 10 good proposals. And these 10 proposals can produce one good project, as shown in Figure 5.2.

*   *   *

"Don't start vast projects with half-vast ideas" Dr. Earle Schumacher, former R&D director at Bell Labs.

*   *   *

Figure 5.2 Project Selection at the Fuzzy Front End

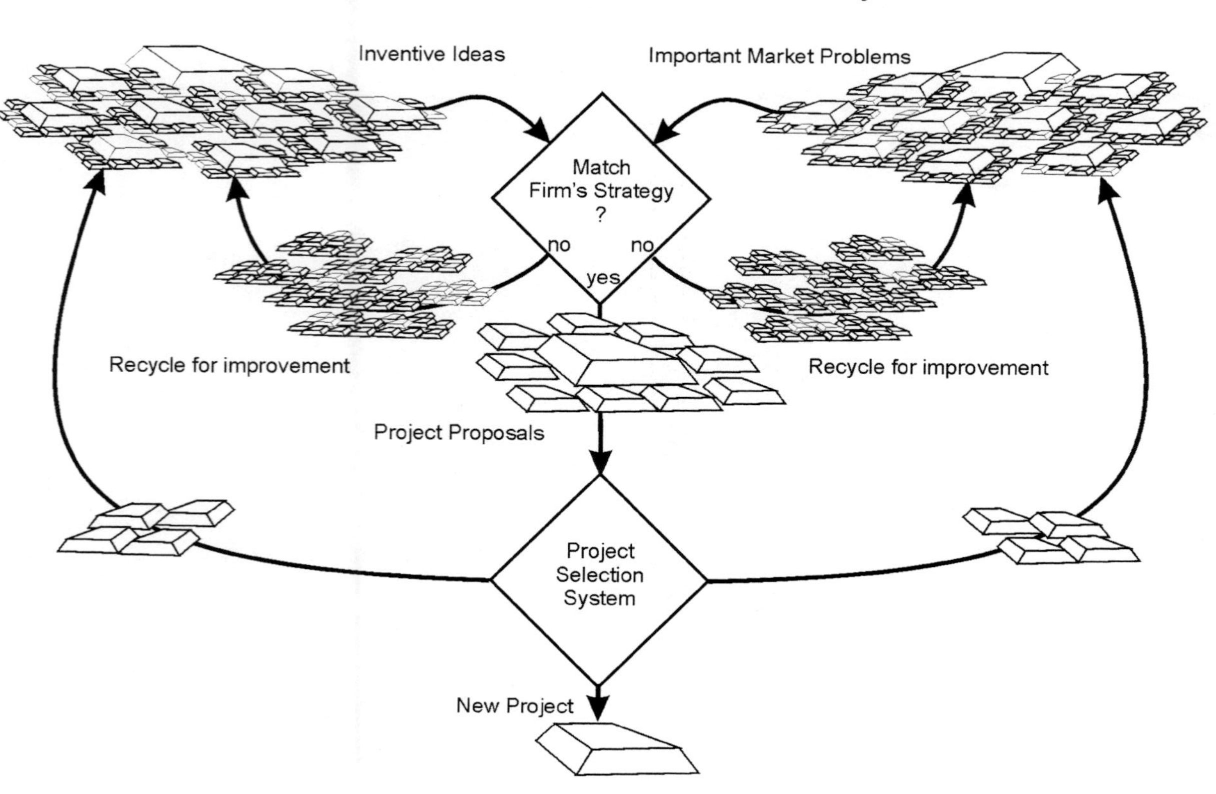

## 5.2 Good Project Selection is Open and Understandable

Once you have a project selection method that fits your company, give a copy of it to every idea champion. It is important that the project selection method be understood and accepted by everyone.[43] This can't happen if the method is a mystery: If the rules are a mystery, it is hard to win. An open project selection system transforms the company's tacit Uncommon Knowledge into explicit Uncommon Knowledge, a competitive advantage.

There are four benefits to having an open process. One, people can score their ideas in private, and so many weak ideas will not be brought forth. Two, it is a lot easier on the ego if a person believes that an idea got a fair evaluation, even though it may not be deemed a winner. Three, the method highlights which parts of an idea are the weakest and so gives direction to where it can be improved. Four, ideas can be evaluated as they come up; they don't have to wait for some scheduled meeting.

Thus, the project selection method can be viewed as identifying which ideas are ready for investment and which are not. This approach avoids loaded words like *killing ideas* and *loser ideas*.[44] Ideas are very fragile at this stage, as are their champions. It is important to keep morale high while some ideas move forward and others do not. A few books have some good suggestions for dealing with the human side of these decisions.[45]

* * *

> Inventive ideas and important market problems are nearly immortal. They rarely die. They are just reincarnated as better ones, over and over. jch

* * *

## 5.3 The People Side of Project Selection

Allocating manpower is the most important task in managing innovation. Furthermore, it is one of the most difficult tasks. Recruit the best people for the projects.[46] That's right, you don't assign the best people; they must be recruited. They may have volunteered already as idea champions. People who are devoted to a program will snatch victory from the jaws of defeat. They will work on it for long hours. They will think about it while sleeping. It is important to recognize that people are not interchangeable parts, especially in innovation.

* * *

> "One factor marked every [new product] failure: Without exception we found we hadn't had a volunteer champion. When we take a look at a product and decide whether to push it or not these days, we've got a new set of criteria. Number one is the presence of a zealous, volunteer champion. After that comes market potential and project economics in a distant second and third." An anonymous Texas Instruments executive.[47]

* * *

All inventors are not equal. Invest more resources in those projects championed by an inventor with high productivity (say more than 1.0 patents per year). Invest less in those championed by less productive inventors. This unequal investment takes insight and courage, but it will greatly improve your overall success rate.

Scores from the project selection system, committed champions, high productivity inventors, and other strengths, will reveal your best projects. Don't starve them. Invest in them for maximum progress until you run out of resources. In doing this, you reduce the cost of your design-in-process inventory.[48]

* * *

Parsimony killed more good programs than design problems.
jch

* * *

Matching technical skills is obvious. Matching personalities is often overlooked. In one of my laboratory manager jobs, I assigned two people to a project who were absolutely ideal technical matches. However, they soon despised each other! No surprise; the project failed.

Let's review the evolution of an inventive idea. There aren't really good statistics on how many good projects go on to be successful new products. And of course, they vary from one situation to another, but we can speak in general terms. The odds against success for ideas are about 3,000:1. When an inventive idea also satisfies an important market need and fits the company's strategic plan, the proposal's odds against success are about 300:1. The proposals that go on to become projects have odds against success of about 60:1. Good projects have odds against success of about 20:1.[49] It's still a long shot, but we're improving the odds at every step.

## 5.4 Summary

The nugget of knowledge in this chapter is that project selection really boils down to five simple issues.

1. What does the customer want that they can't get from anyone else?
2. Are there lots of customers eager to buy? Is it a head-slap?
3. Can we do it better than anyone else? Can we protect it from competitors?
4. Does it fit our company?
5. Have we got the right people on the team?

# 6

# *Mining for Successful New Products*

So now you have several projects that appear to be good candidates for new products. Now the work should focus on reducing the uncertainties in the projects.[50] First and foremost, we want to ensure that the project's goals are correctly defined. Then we want to reduce the range of probabilities for the most important outcomes.

## 6.1 Managing the Project with a Stage-Gate System

Successfully managing a project requires an effective Stage-Gate system, as shown in Figure 6.1. The Stage-Gate system helps structure the problems to be solved, both by time sequence and by importance. There are many good descriptions of Stage-Gate systems and we will not duplicate them here.[51]

Figure 6.1 Stage-Gate Project Development System

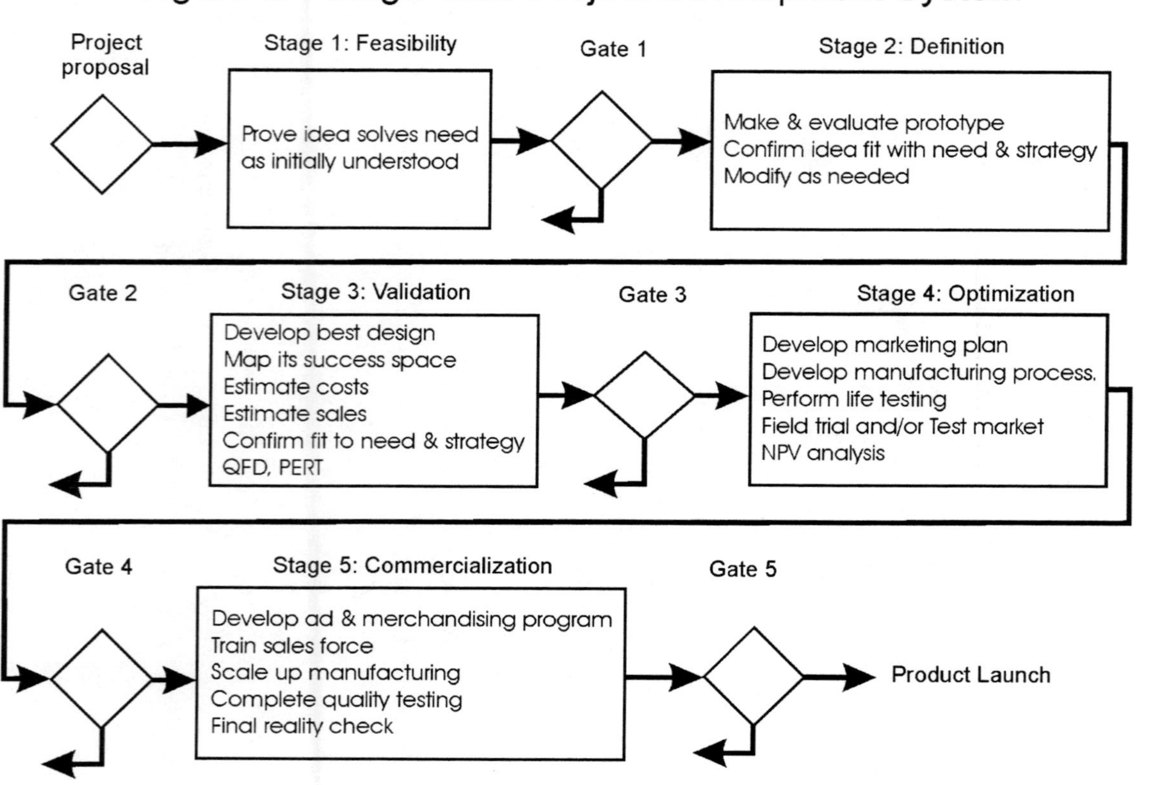

Stage-Gate systems have been widely used for at least thirty years. You might think that they have improved the odds of new product success. But the evidence is weak, and not very convincing.[52] One reason is that all companies have benefited from this Common Knowledge and so none have gained a competitive advantage. Another reason is that effective Stage-Gate implementation has a relative small effect on new product success. One of the objectives of this book is to improve the quality of projects that go into the Stage-Gate system, which is recognized as a needed improvement.[53]

The Stage-Gate system shown here assumes that the initial screening has already been completed, as discussed in the preceding chapter. This Stage-Gate system also places more explicit emphasis on using prototypes as the tool for developing an accurate definition of the fit among the inventive idea, important market need, and good strategic statement before using quantitative methods.

Simply put, a Stage-Gate system is a guidebook for the project. It breaks the journey into stages, with specific problems to be addressed and solved in each stage. At the end of each stage is a go/no-go decision. That is, has the project solved the problems associated with the stage? If it has, then it can be approved to enter the next stage. This usually means additional investment in manpower, equipment, or market research. If the project has not solved the problems, then it does not advance to the next stage. Sometimes the project must retreat to an earlier stage. Sometimes the problem is fatal and the project is cancelled.

One of the principal advantages of the Stage-Gate process is that everyone knows what is required at the next Gate. So the project leader or the laboratory manager can take steps to remedy a lagging part of the project in time. I've worked with Stage-Gate processes for thirty years and the go/no-go decision is rarely simple. Usually, some problem is not completely solved. However, that problem may not be great enough to hold up progress in all the other areas. This is where experienced management is key, because the go/no-go decision usually is made by the company's executive committee (the heads of laboratory, marketing, manufacturing, and finance). Because these people are busy, the decision is made in a regularly scheduled meeting, perhaps quarterly. It would be a waste of resources to have part of the project team wait several months while another part solves a stubborn problem.

Of course, the Stage-Gate system for each company should be different. It should reflect its strategies and tactics, strengths and weaknesses, and threats and opportunities. When tailored to fit, the Stage-Gate system transforms the company's tacit Uncommon Knowledge into explicit Uncommon Knowledge, creating a competitive advantage.

## 6.2 Relationship Between the Project Team and the Manager

The kinds of talents needed in each stage are generally different. In the first stage, the key talents are the ability to perceive the greatest uncertainties, generate successive inventive solutions, be energetic, and be determined.[54] More specifically, the full-time team members are usually an inventor, a junior engineer or scientist, and a technician. Even though marketing and manufacturing might work only part-time on the project, they must be kept informed on the project progress at least weekly. It is vital to establish mutual trust and respect in the beginning. There will be many obstacles and setbacks during the project. A healthy level of communication is vital to having confidence they can be overcome. Otherwise, the team will drift apart, make unilateral decisions, and ultimately engage in conflict. Once conflict begins, it is very hard to establish trust.

Once the project is staffed and the tasks for the first stage are known, the project team begins work. One of the hardest things for a laboratory manager is to give the team the freedom to run the show. Micro-managing all the projects in a laboratory spreads the manager too thin and diminishes the team's enthusiasm and commitment. Put another way, when the inevitable setback happens, a micro-managed team will blame the manager and not take ownership of the problem. The project team will build enthusiasm and commitment when they are making their own decisions. This is empowerment.

* * *

> Participation is the mother of commitment. Commitment is the father of success. jch

* * *

The classical list of a manager's tasks is: organize, communicate, control. This model may be adequate when workers are interchangeable. But it is not effective when workers need to identify and solve problems on their own.

* * *

> The Myth of Control: Control is the minimum performance of management. Especially for managing innovation, coaching and encouragement are far more important. jch

* * *

Of course, it would be foolish to cast them completely adrift. The laboratory manager has a responsibility to see that they are working on the key problems.[55] Frequent, informal conversations with the project leader are an important way of monitoring this kind of progress. If the conversations are in the style of "How can I help you?" they are not likely to be perceived as micro-managing. Put your resources where your mouth is; walk the talk. Assign an extra technician temporarily. Suggest an expert who might offer special advice. You cannot kill a project with kindness, but you can kill it with neglect.

## 6.3 Enduring Setbacks

Almost every innovative project encounters obstacles and setbacks, because of its built-in uncertainties. If the project is an innovation, you haven't done quite the same thing before. This is one reason why formal project planning is not very useful for innovations. It is nearly impossible to estimate how long a task will take if you have never done it before.

The problem-as-given is accurate only in the artificial world of academia. In the real world, the market problem will probably have missing information, vague information, and even errors. The inventive solution will probably not work on the first try. Sometimes the original idea must be discarded and a new one must replace it. These situations are normal and should be expected. Remember, the purpose of the early stages of new product development is to reduce uncertainty.

To start with, we want to be sure that we are addressing the right questions, rather than getting the right answers. Then we want to find ways to reduce the range of probabilities of outcomes. One of the hardest things for a team to accept is that the experiments should be designed such that there is a reasonable chance for failure.[56] Yes, failure! New product development is a journey of discovery. Experiments help us create the map and find the boundaries of the success space, as shown in Figure 6.2.[57] The bigger the space, the better the product will be. If all the experiments are designed for success, we never find the boundaries.[58]

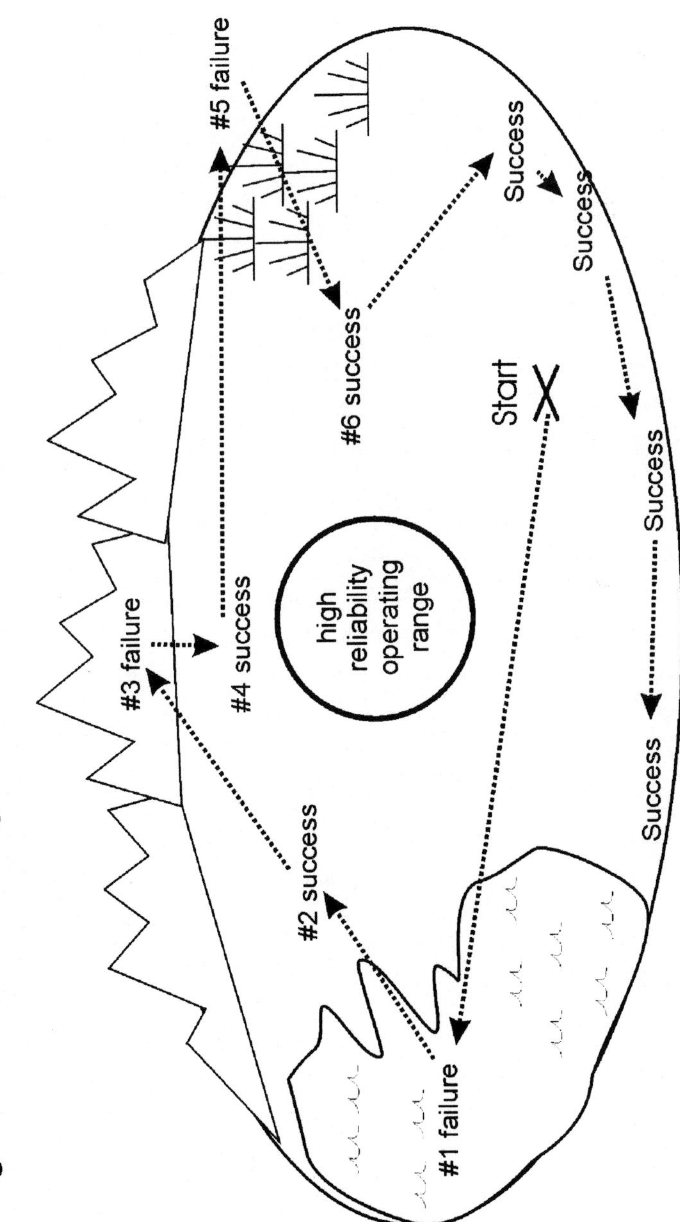


Figure 6.2 Discovering the Boundaries of the Success Space

Let's be clear about this kind of failure. It is not a failure in the customer's hands. By mapping these experimental failures in the design space, we also map the high reliability operating range. Thus, forcing failures in the design stage is necessary to minimize failures when the customer uses the product.

* * *

> "If everything you're trying works, you're not trying hard enough." Gordon Moore, co-founder of Intel, leader in microprocessors, and author of Moore's Law, the vision and productivity engine of the last 20 years.

* * *

An example may help make the point. A project team in my lab was developing a new electrical connector for telephone wiring, principally intended for developing countries. The international marketer was concerned that the connector wasn't rugged enough to stand up to mistreatment. The project leader assured him that it was, but the marketer remained skeptical. As lab manager, I suggested a trial-by-torture. The project leader said, "He'll break it because he wants to." However, I pointed out that the international marketer might be right and it would be best to find out before we spent a million dollars on the manufacturing equipment. So we set up several hundred prototype connectors and had a dozen people work with them. Naturally, the project leader had no failures. Most of the other people had very few failures. However, the international marketing manager did everything he could to force failures and about half of his connectors were broken. His conclusion was, "If I can't break more than that, the product is rugged enough." He was a champion for the product ever after.

* * *

"The biggest job we have is to teach a new hire how to fail intelligently." Charles F. Kettering, long-time R&D Director of General Motors and inventor of the self-starter and other automobile breakthroughs.

* * *

It is important to distinguish between a failure and a mistake. A failure is an event that did not produce the desired result. Failures are an integral part of learning and a natural consequence of experimentation and exploration. The only learning without the possibility of failure is that which is based on Common Knowledge; it can prepare you to participate, but it cannot help you to excel. Mistakes are failures that could have been prevented. Thus, a mistake happens when some Common Knowledge was not known or properly evaluated. Mistakes are failures that many other people would have made into a success. On rare occasions, what appears to be a Mistake is actually a repeal of Common Knowledge; these repeals then become Uncommon Knowledge, and sometimes Breakthrough Opportunities.

You can create Competitive Advantage only by creating Uncommon Knowledge. If you obtain the desired results, we call it success. But in discovering Uncommon Knowledge, the outcome is not certain and therefore, the outcome might be undesired results, and thus failures. The only failures that must be avoided are the ones you cannot survive or learn from. When you survive and learn from failure, you are gaining more Uncommon Knowledge that improves your competitive advantage and opportunity for success. These relationships are shown in Figure 6.3.

* * *

"We stumble on many of our best discoveries." William E. Coyne, Sr. Vice President for Research and Development, 3M Company in the 1990s.[59]

*                *                *

Figure 6.3 Competitive Advantage, Failure and Mistake

| Result of Experiment | Knowledge Used: Common | Knowledge Used: Uncommon |
|---|---|---|
| Desired | Mediocrity | Competitive Advantage |
| Undesired | Mistake or Breakthrough Opportunity | Failure |

There are four kinds of "facts:" happy and true, happy but false, sad and true, and sad but false. It is human nature to challenge those sad "facts" that contradict our expectations. However, the most dangerous "facts" are those that we happily believe but are actually false. Thus, it is always important to test even the things that fit our expectations. For example, the repeal of Common Knowledge discussed above is a "fact" that is false.

## 6.4 The People Side of Projects

It is important for marketing to be patient with these setbacks. In these early stages, the level of mutual trust and respect between marketing and the lab is begun. The importance of harmonious relationships between these two groups has been proved over and over.[60] The tables will be turned on marketing when the customer evaluates the prototype and changes the problem, forcing a change in the design.

One of your key responsibilities is to see that the team is working harmoniously. Are there signs of mutual trust and respect? This is best detected by management-by-walking-around. Observe the people at work. Are they enthusiastic? People engaged in distrust and disrespect are rarely enthusiastic. It is probably best if the laboratory manger does not attend all the team meetings; it works against empowerment. However, do the team members have lunch together? If not every day, at least every week.

It is not a bad idea for the laboratory manager to have lunch with the part-time team members, such as the marketer, once in a while. Keep the atmosphere informal. A mild, "How's it going?" is a good opener. Watch for enthusiasm. Listen for head-slap versus foot-shuffle. However, don't try to be a psychologist. If you see signs of trouble, put the problem to the team leader. Coach the leader to understand that conflict mostly grows with time and eventually will kill the project. A team leader doesn't need that kind of blemish on his or her career.[61]

* * *

> "The plan is nothing. Planning is everything." Dwight D. Eisenhower, describing the invasion of Europe. Put another way, when a group of people engage in planning, they integrate all their Uncommon Knowledge and create their Competitive Advantage. With this foundation, they can readily make changes as situations evolve. Without this integration, a plan is a house of cards; ready to collapse at the first setback.

* * *

## 6.5 Prototypes Help Get the Definition Right

The Stage-Gate system is a process, not a solution. Evidence toward finding solutions to the problems comes mostly from prototypes. Make prototypes early and often.[62]

A prototype is a kind of language that lets the customer and innovator communicate. A little more discussion will make this clearer. Music is the language of feelings. Writing is the language of description. Numbers are the language of accounting and business. Drawing is the language of design. The language for one field does a poor job in another field. Don't use words, numbers or drawings alone to communicate innovation. Nothing does it better than a prototype. Prototypes capture explicit and tacit knowledge from both the customer and the innovator. Prototypes help get the definition right.

Show the prototypes to the marketer. Listen to his feedback. When he or she is satisfied, show them to an insightful customer. Listen carefully to their feedback. Human nature tends to keep people from giving bad news in this situation. Watch for head-slap versus foot-shuffle. Be prepared to ask some open-ended questions, such as "How could we make it better? Tell me how this is important to you." Listen to what they don't say. Never ignore bad news; it won't go away.

* * *

> Remember the Edsel: Customers can change their minds while you are working out the solution. jch

* * *

When evaluating prototypes the head-slap versus foot-shuffle law becomes even more important. When customers first see a head-slap prototype, its correctness, usefulness, and uniqueness are immediately apparent, like a revelation. They are overcome with a lust to possess the prototype. They are reluctant to give it back to you. Price is rarely mentioned. On the other hand, when customers first see a foot-shuffle prototype, its attributes appear universally suspect, even flawed. The purchase decision is subjected to thorough analysis. Price is one of the very first issues discussed. The idea may be new and useful, but it is so close to the existing methods that it isn't obvious to the buyer that having it is worth the inconvenience of taking on any change.

* * *

Tower of Pisa Law: Weak foundations destroy good design.
jch

* * *

Even if the responses are "Wow!" quite likely the customer will want changes after evaluating the prototype. Quite often the solution changes the problem. It is natural to feel frustrated by this kind of setback. However, it is a natural part of creating Uncommon Knowledge. Your competitors might have discovered this important market problem. They might even have made prototypes. But they probably have not discovered the necessary changes. Thus, your company has a better level of Uncommon Knowledge. Usually, the people who are the most frustrated are the lab people. This is when mutual trust and respect is really needed to avoid everyone's blaming someone else.

Here's a real-life example of a customer changing the problem. One of the side-effects of cancer treatment is loss of red blood cells. Low production of red blood cells may indicate the treatment is too severe for that patient. New red blood cells contain a structure that looks like a pile of spaghetti. These can be seen under a microscope and counted by a technician. But this manual process is tedious, time-consuming, and sometimes inaccurate. Our first prototype could reliably count these structures. But the second round of tests had counts that were too large. When questioned, the customer said, "Oh, many cancer patients have Howell-Jolley bodies in their blood cells. I guess the first batch of samples didn't have those bodies. But any useful instrument must not count the Howell-Jolley bodies." The Howell-Jolley bodies are a structure that looks like a small dot. Our technology couldn't distinguish between the pile of spaghetti and the Howell-Jolley body. The project cancellation was no one's fault; such events are an inherent part of getting the project's definition right.

Afterwards call the customers who examined your prototype and thank them for their interest. This is a great chance to say, "We're always interested in your ideas about important market problems." They could be the source of your next successful program. If you get positive feedback, show the prototype at the next Stage-Gate meeting. The executive committee gets tired of seeing charts and graphs. They love touching something real. A prototype does wonders for the team's credibility.

Two things can go wrong with showing prototypes outside the company. The first is that a competitor might learn about what you are doing and catch up with your project. The second is a patent issue called divulgation. For many countries, divulging the invention outside the company can seriously damage your opportunity for a patent. In many cases, a confidential agreement with the customer can solve both these problems. This is an area of patent law that can change rapidly. Your patent counsel can guide you with the latest information.

## 6.6 Product Development: Validation, Optimization, and Commercialization

In the middle ground of the Stage-Gate system, the project has passed most of the major hurdles. But these are the rare projects. About 90% will have been recycled and the remaining 10% are the strong contenders. But the odds are still about ten to one that even a strong contender will become a successful new product.[63]

As the project moves along through the various stages, the marketing plans and product descriptions become more and more detailed. The volume of prose and charts can create an illusion of low risk. The most important goal of these documents is to communicate among the functions, to create the consensus view, and to describe the project's Uncommon Knowledge. Without this, mere tonnage is waste.

* * *

> "One qualitative tool for providing focus is so important that it deserves a separate discussion. It is the product mission, or what some would call the product's value proposition. This is the answer to the question, Why should a customer buy this product instead of the competitor's? If this question cannot be answered in a compelling way in twenty-five words, then there is a fundamental problem with the design of the product [or more-likely, the market opportunity]."[64]

* * *

These final stages are also when the more quantitative project management methods can be used effectively. One of the best methods is Quality Function Deployment, with several good resources in the endnotes.[65] But before the customer has evaluated prototypes, you don't know much about "the voice of the customer" that is an important part of Quality Function Deployment and other rigorous methods of getting the definition right.

These final stages are also where project scheduling charts are valuable.[66] In the early stages, there are many setbacks, some of which will change entirely the downstream activities. I once observed an early-stage project that devoted a full-time person to managing and revising its scheduling chart. The chart was revised so often that it became the source of endless less-than-humorous comments. It would have been better to invest that full-time position in helping solve the setbacks.

* * *

> Sonia's Law: An innovation is like a baby. The conception is more fun than the delivery. jch

* * *

In the final stages, all of the major uncertainties will have been resolved. That is, the project team will have determined what the customer needs and whether the product will satisfy that need. This is the time to evaluate projects based on variations of net-present-value calculations.[67] Usually, this calculation will be done by your accounting department. The key component of these calculations is a sales forecast. This is a very difficult task, because the marketer must assume a market size and growth, a market share size and growth, a unit price value and any changes over time. These estimates contain assumptions about the responses of competitors. There are many responses, but the most obvious will be price cuts. One thing is fairly certain, competitors will not lose without a fight.[68] I once examined over a hundred new products' sales for the third year after they were launched. None of them were within 20% of their forecasted sales.[69] Most were either a factor of two larger or a factor of two smaller. If you think that this deviation was because the marketers were lazy or stupid, try this experiment: Forecast the winners of the NFL Super Bowl for the next five years and the exact scores. It's the same problem, just a different playing field.

The final stage is launching the new product. Sometimes the need is so urgent, everyone is united in pushing it out the door. However, sometimes one function or another keeps finding improvements and refinements. Sometimes the inventor wants improved versions. Sometimes the marketer wants more test markets. Sometimes manufacturing wants more time to improve quality. There is a time to call all these to a halt. They can be part of the second generation. You'll know when to make this decision; the customer you've shown the prototypes, or engaged in a field trial, or participated in a test market, will be clamoring for delivery. If they aren't, you may have deluded yourself; what you thought was a head-slap may be really a foot-shuffle.

* * *

Perfection is the enemy of progress.

* * *

Figure 6.4 Survival of Ideas

| Ideas | Stage | % Survive |
|---|---|---|
| 3,000 | Ideas | |
| 300 | Proposals | 10 |
| 60 | Projects | 20 |
| 9 | Definition | 15 |
| 4 | Validation | 44 |
| 2 | Commercialization | 50 |
| 1 | Success | 50 |

The main purpose of the Stage-Gate System is to reduce uncertainty. The survival prospects of an idea improve with every stage, as shown in Figure 6.4.

When you've launched the new product and feel that you've done everything right, you are entitled to congratulate yourself. Bask in the warm glow of its achievement. Then jump in the lake of cold reality. The odds are only even that the product will be a success in the marketplace.[70]

## 6.7 Summary

The nuggets of knowledge in this is chapter boil down to five simple commandments of new product development.

1. The customer is always right—especially when they know what they are talking about. Always ask checking questions.
2. Both marketing and lab people are responsible for assuring the program has value in the customer's eyes.
3. All real-world problem statements have errors, ambiguities, and missing information. The solution to the problem helps create an accurate definition. Early prototypes are the best tool for communication.
4. Winning programs are both fast and correct. It is not always necessary to compromise between speed and accuracy.
5. Never withhold bad news. Every project has setbacks. Management's job is to help the team overcome setbacks.

# 7

# *Mining for Improved Future Innovations*

By now, you've got several promising new product development projects well on their way to success. Probably some of the existing projects that had been well along their course have been born as new products and are launched into the marketplace. This is the time for some intelligent planning.

## 7.1 Invest Wisely in Product Improvement

First, do a good job of investing in product improvements and product support. Don't let the existing products be orphans, but don't let the teams get bogged down in customer training or endless improvements. If the team has done a good job in the Stage-Gate process, the other functions will do their part, such as training. In my experience, many requests for minor product improvements are really a sales issue. Good sales reps can sell around minor shortcomings.

Second, it is important to cast a critical eye on existing products. Are there substantial defects in some existing products? Have competitors introduced products that are substantially better? Is there a logical next generation for some existing products? Are supplemental products needed? These product improvements are easier to evaluate than innovations. Usually, there are few uncertainties. The product specifications will be clearer. The sales and costs can be estimated. Standard net-present-value methods can be used.

However, the very fact that there is so much low-risk and quantitative information can be deceiving. In my experience, there remains the substantial uncertainty of the level of customer commitment. I once tracked 100 sales requests for product improvements. We agreed to tackle over half of them. About half of these actually had a prototype made and shown to the customer. However, none of them generated enough response even to become project proposals. From this vantage point, the best product development is no product development. It has zero cost and infinite speed.

* * *

> "The temptation in the existing business is always to feed yesterday and starve tomorrow." Peter Drucker, author of many books on management

* * *

In many ways, deciding the allocation of manpower resources between product improvements and innovation is similar to allocating personal investments between stocks and bonds. It may seem safer to invest entirely in product improvements or bonds, but you risk losing value due to inflation or innovation by competitors. Conversely, investing entirely in innovations or stocks may produce excessive fluctuations in sales or portfolio value. Experienced executives know that failing to innovate is riskier than innovating intelligently.[71]

## 7.2 Improving the Project Selection System

Having dealt with the easier, but important, issues of product maintenance, it is now time to address more innovations. If this is your company's first experience with the project selection system, you've probably discovered some useful refinements. A project selection system should be a living process, ever evolving and strengthening as you gain a better understanding of important market needs, inventive capabilities, and even changes in the company's strategy. Some of your recent new products have succeeded and some have failed. All this is Uncommon Knowledge and should be incorporated into the project selection system to improve your company's competitive advantage. It may also be valuable to document this Uncommon Knowledge in a lessons-learned database. There are several good resources on this topic.[72]

If you've done your work properly, the project selection system has gained the confidence of marketers and inventors alike. Some of the project proposals that needed more work have been improved. Old ideas have been revised. New ideas have been spawned. It is time to begin again. Don't be passive. Ask the marketers, "Where's the next important market problem?" Ask the inventors, "Any ideas cooking?"

## 7.3 Improving the People Side of Innovation

This is also a good time for evaluating the people side of innovation. If your company is going to have a stream of successful innovative products, it needs a stream of people with fresh ideas. Now we are mining for human nuggets. By now you've also learned more about which of the lab people are inventive and which are not. Recalling Chapter 3, only about 36% of the people in a lab will have at least one patent. And only 10% of inventors have at least five patents, but these people produce half the patents. That means that most of the people in a laboratory will never produce a patent. Returning to our example of a lab with 35 people, only one of them will be a prolific inventor with at least five patents, as depicted in Figure 7.1.

Figure 7.1 Number of Patents Held

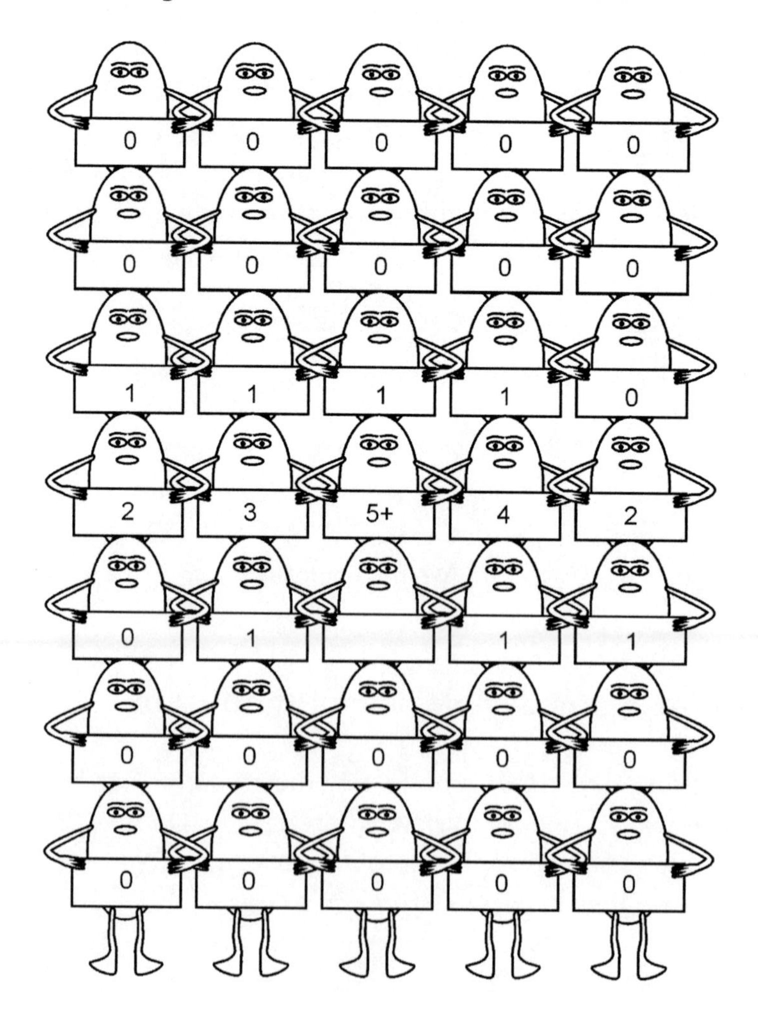

While you are evaluating people for their inventive potential, keep in mind that their outputs will fluctuate over time. Furthermore, they fluctuate quite differently from what you might expect. In Appendix C we find that the vast majority of inventors have a time pattern of patents that follows the Poisson distribution. This means that they will have more years with below average outputs than years above average. For example, let's examine an "average" inventor with 0.7 patents per year. That inventor's yearly distribution is shown in Figure 7.2. Thus, he or she has even odds of having zero patents in any given year. This apparent slump is not caused by their failure but is inherent in the inventive process. These low production years are offset by years with much larger than average production. This example is for an "average" inventor. Recall from Chapter 3 that the distribution of inventive productivity is exponential. Thus, most of the people in a lab have productivity that is below average. Their distribution will have even more years with zero outputs. The important conclusion is that you should not base your opinion of someone's inventive potential on his or her results for just one year.

Figure 7.2 Poisson Distribution of Patents

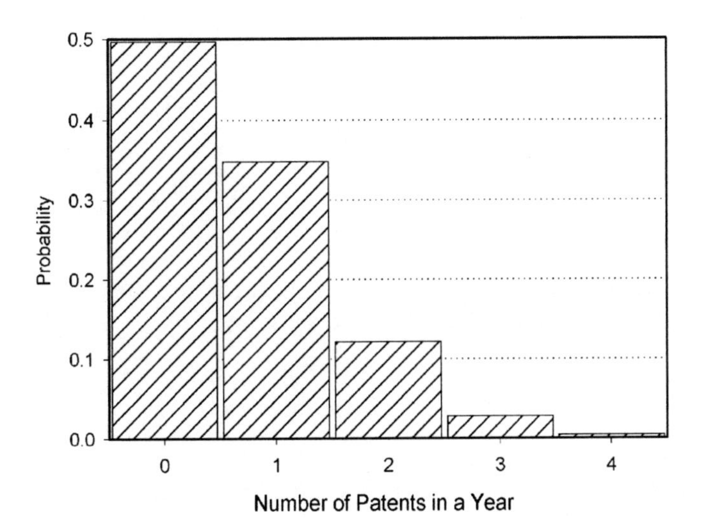

Inventors need at least five patents over at least three years to demonstrate a useful estimate of their average productivity. When you calculate an inventor's patent productivity, be sure to use the correct equation, which is ((number of patents)-1)/(LastYr-FirstYr).[73] Here, LastYr and FirstYr are the dates of the last and first patents and are continuous variables. For example, a patent that issued on October 1, 2000 would have a value of 2000.75. As shown in Appendix C, estimating patent productivity is limited to the top 10% of inventors.

Thus, the vast majority of people in a typical lab have little inventive potential. Some of these less-inventive people will have more success in another function.[74] Some may be better at manufacturing process engineering than at product development. Some may be better at product training. There are many jobs in your company that need technical knowledge, not just those in product development. Similar to a sports coach, a laboratory manager has a responsibility to place people in positions that take advantage of their strengths. If these transfers are done with grace and sensitivity, you'll often find the individual thanking you for your guidance.

* * *

> To get the highest performance, find out what each person does best, and help them do it. jch

* * *

There is an obvious criticism to this outplacement strategy. Why not train these less-inventive people to be more inventive? This question assumes that there is an effective training program. So far, none has proven substantially effective, though many have been proposed over the last 40 years.[75] Furthermore, inventors and scientists themselves are highly skeptical about training versus innate talents.[76] Indeed, the psychologists themselves admit to this lack of evidence.[77] On a more positive note, some eminent scientists have described lists of techniques for solving difficult problems, as listed in the endnotes.[78] These can be useful exercises for inventors and aspiring inventors at every level.

Transferring people out of the lab opens positions for new people. Use the Inventor Profile, described in Appendix B, to identify new hires or existing workers who have the potential to become prolific inventors. One of the Profile's greatest values is destroying myths. Most significantly, an exceptionally high undergraduate GPA is not an indicator of higher inventive productivity. Neither is broad interests or being left-handed. The best predictors are hard work and an advanced degree with a high GPA. The Profile is a useful tool, but it is only a supplement to your experience and judgment.

One important predictor of inventive talent cannot be captured in an interview or on a test: Successful inventors have a healthy discontent with the status quo.[79] When they see a product for the first time, they scour it for weaknesses. In their view of the world, everything ought to be replaced, even their own inventions. You can assess this characteristic best by watching them work. However, this chronic discontent can have its dark side. Sometimes it means you have to coach them to let an invention go and proceed to the next great idea.

* * *

> "If your favorite tool is a hammer, every problem looks like a nail." Mark Twain

* * *

It is common for companies to hire people who have some expertise in the company's key technologies. I suggest that this may not be a best practice. The reason is that the company should already be a master of all the Common Knowledge related to its technologies. Therefore, it would be rare for a new graduate to possess Uncommon Knowledge about the company's technologies. So this kind of new hire adds no new knowledge. It is wiser to hire people who have some expertise at the periphery of the company's key technologies. These new hires can bring pre-packaged and crystallized Common Knowledge and perhaps some Uncommon Knowledge. Such hires can expand the company's success space. Even if the new knowledge doesn't produce high impact new products, the company has eliminated a blind alley. This is Uncommon Knowledge also.

In addition, character traits are more important for inventive productivity than knowledge, skill or experience. Given a reasonable intelligence, knowledge can be taught, skills can be learned, and experience can be gained. But without inherent traits of honesty, creativity, motivation, focus, and cooperation, all the knowledge, skill and experience cannot be put to good use. It becomes a roaring engine with no steering, more of a hazard than a help.

Always remember: Only people make innovations. A company's structure, policies, procedures, and even the project selection system do not make innovations. These are inanimate (and some might say inhumane) devices; they may help or hinder innovation. Moreover, the Inventor Profile indicates that they do not have a major impact. It is the people, their talents, mutual trust, and respect, who make innovations. The lab manager must encourage them, coach them, and praise them. It is also important to recognize that it is human nature to respond to opportunity. If your company has a culture that clearly says "Innovation is expected here" you'll find that innovations will be plentiful and successful.[80] This factor has been the most frequently mentioned by both experienced managers and academic researchers, previous to the Inventor Profile.[81]

Current practices for managing innovation are far, far from being truly effective. First, we can compare inventive productivity across companies. One study found there is only a small difference in the overall patent productivity of companies and no difference in distribution of the most productive inventors.[82] Second, we can compare the productivity of inventors to other professions using the Performance Improvement Potential (PIP) method.[83] This method examines the ratio of the highest producer to the average. Amongst our candidates for the highest producer are Thomas Edison at 22 patents per year, Jerome Lemelson at 12, Michael D. Rostoker at 26, and Mark G. Benz at 8. We choose an intermediate level of 10 patents per year. The average inventive productivity for large, industrial companies is about 0.7 patents per year. [84] Thus, inventors have a PIP=10/0.7=14. This is a value similar to other professions with a large component of tacit knowledge, such as sales, or teaching. However, Performance Improvement Potential in professional athletics, which also has a large component of tacit knowledge, are generally less than 1.5. This difference between inventors and athletes tends to indicate that there is substantial room for improvement in the management/coaching of innovation.

* * *

> "The greatest secret in the world is that you only have to be a small, measurable amount better than mediocrity…and you've got it made." Og Mandino, author of best selling self-help books.[85]

* * *

## 7.4 Summary

The nuggets of knowledge in this chapter is that you need to apply what you have learned about your recent projects. You can create new Uncommon Knowledge. This will give your company even stronger competitive advantage.

# 8

# *Mining for Better Management of Innovation*

There are many managers of innovation. First, there are the inventors and aspiring inventors who want to manage their ideas better. Second, there are the first-level managers in product development labs. Most of them are about average. But you want to be better than average. You want to be a nugget yourself. You want to be the best at managing innovation.

However, innovation is just a part of the overall work in a product development laboratory. Thus, we have not discussed a broad range of important topics, such as budgets, organizational structure, and performance reviews. There are many good books on this subject listed in Appendix A.

The previous chapters focused on how you, as the manager of innovation, can improve on the processes and people around you. All these are good and useful tools. However, they depend upon two important character traits of the innovation manager—leadership and courage. Put another way, ordinary managers might be able to get by with just being an administrator—an obedient implementer of the company's policies and procedures. But managing innovation is different. This

self-improvement applies to inventors and aspiring inventors as much as it does to laboratory managers.

## 8.1 Becoming a Leader

As we observed in Chapter 4, most companies do not have a clear vision of what their new product strategy should be. Even the companies that do have a clear strategic statement cannot lay out the innovation manager's job in clear terms; all the uncertainties make it impossible. Thus the manager of innovation must be a leader. He or she must articulate a clear vision of the future that others will support. Being a leader is much, much more than just conjuring up a vision of the future and ordering everyone else to follow.[86] Leaders are not self-made; they are elected by their constituents. Notice that I didn't use the word "followers," because it implies a passive, subordinate role. Constituents are the power base of the leader; without them, a leader is irrelevant. Thus, a leader must constantly earn the credibility of his or her constituents.[87] The good news is that, leadership can be learned and improved with resources in the endnotes.[88]

## 8.2 Developing Courage

Managing innovation means that decisions must be made before all the facts are clear, as discussed elsewhere in this book. These decisions require the manager not only to be a leader with vision, but also to have the courage to implement that vision. Indeed, this courage takes many forms.

- The courage to focus effort on opportunities rather than problems.
- The courage to invest in projects before all the facts make the choice clear to competitors.

- The courage to postpone projects before all the facts make the choice clear.
- The courage to endure a few failures if they create Uncommon Knowledge.
- The courage to allocate resources unevenly.
- The courage to say Yes and No unequivocally.
- The courage to outplace less-inventive employees.
- The courage to write a clear, concise and motivating strategic statement.
- The courage to decide which customers, markets and technologies will not be of interest.

In this book, as in most other books on innovation, the discussion has focused on improving innovation. You might get the idea that everyone is in favor of innovation. This is untrue. It is an unfortunate fact of life that many people fear of innovation. Those people who are the company's keepers of policies, standards, and measures of consistency often perceive innovation as an unnecessary distraction from the smooth, predictable path they earnestly desire. You will find them commonly in accounting or manufacturing, where consistency and accuracy are central cultural beliefs. Often, these people will seize every setback as evidence of mistake rather than the necessary learning to gain competitive advantage. These people often remain blithely unaware that relying on Common Knowledge and its predictability almost guarantees eventual destruction by some innovative competitor. In my experience, it is important not to fight these people; they occupy powerful positions in the company. The best way to avoid inflaming their fears is to minimize the impact of setbacks. That is, when a setback occurs, avoid the words *mistake* or *failure*. Describe the remedy. Remind them that the goal is to have a competitive advantage that will earn more profits (making the accountant happy) and have large production (making the manufacturing people happy). Lastly, this is where courage is important, as well. Put a smile on your face, act confident, stick to solving problems, even while you suffer the mud-slinging. This requires leadership and courage.

As the manager of innovation, you must be more than just an administrator. You must exhibit leadership and courage. Leadership can be learned. Courage comes from a successful track record.

## 8.3 Learning More About Managing Innovation

There are many hundreds of books and many thousands of articles related to innovation. Obviously, a busy manager of innovation cannot read them all. How can a busy manager of innovation even decide which ones to read? In the past five years, I had the time to do this reading. In the sections below I give you my opinions. I'm sure that some others will disagree. It is hard to draw fine distinctions amongst the best.

## 8.4 The Top 1% Books

For the busy manager of innovation, I have chosen what I think are the best books to help you improve your performance and are all in stock at the major bookstores. They are the nuggets; the best seven of the 700 books I've read related to innovation. First, they are listed in alphabetical order. Below is a table showing where their content supplements this book. If you want to read more, Appendix A lists books that I think are the top 10% of books about managing innovation.

Cooper, Robert G. 1993. *Winning at new products (2nd ed).* New York: Perseus Books.

Among all the books on new product development, this one and the next have the firm foundation of real-world experience across many companies and industries. This foundation provides useful lists of factors for success and failure. This book is also the most comprehensive about the new product development process. However, the NewProd system is improved in the next book.

Cooper, Robert G., Edgett, Scott J., & Kleinschmidt, Elko J. 1998. *Portfolio management for new products.* New York: Perseus Books.

The main advance of this book over the one above is the improvement in the NewProd project selection system by using anchoring phrases. This is an important enhancement, leading to a better scoring system.

Davidow, William H. 1986. *Marketing high technology: An insider's view.* New York: The Free Press.

The title and content of this book stress marketing, but much of it applies to creating clear strategic statements. It focuses on knowing the company's strengths and applying them against a competitor's weaknesses. Much of the content is based on the author's real-world experience in making Intel the global leader in microprocessors.

Drucker, Peter F. 1985. *Innovation and entrepreneurship.* New York: Harper & Row.

This is the first book to present innovation in its entirety and in a systematic form. It has not been surpassed in its eloquent explanation of the key elements of innovation: why companies should innovate; that innovation is mostly good; that setbacks are an expected part of innovation.

Reinertsen, Donald G. 1997. *Managing the design factory: A product developer's toolkit.* New York: The Free Press.

This book has the most specific, practical advice for the new product development process. For the manager of projects already in development, this is an excellent book. The chapter "Get the Product Specification Right" applies mostly to Stage 2: Definition of our Stage-Gate system. The chapter "Manage Uncertainty and Risk" applies to our Stages 3 through 5, namely Validation, Optimization, and Commercialization.

Schrage, Michael. 2000. *Serious play: How the world's best companies simulate to innovate.* New York: Harvard Business School Press.

This book is an exhaustive advocate of prototyping—why, when, where, and how. It is especially good for the many ways that prototypes should be evaluated.

Wheelwright, Steven C., & Clark, Kim B. 1992. *Revolutionizing product development: Quantum leaps in speed, efficiency, and quality.* New York: The Free Press.

This book's focus is on Stage 5, commercialization. It has many useful diagrams that help to manage projects. It describes the important task of cross-functional integration, the development of mutual trust and respect. It also contains a good discussion of the importance of prototypes.

The books described above support some of the chapters in this book. The first row in Figure 8.1 is the chapter number.

Figure 8.1
Top 1% Book's Support of this Book

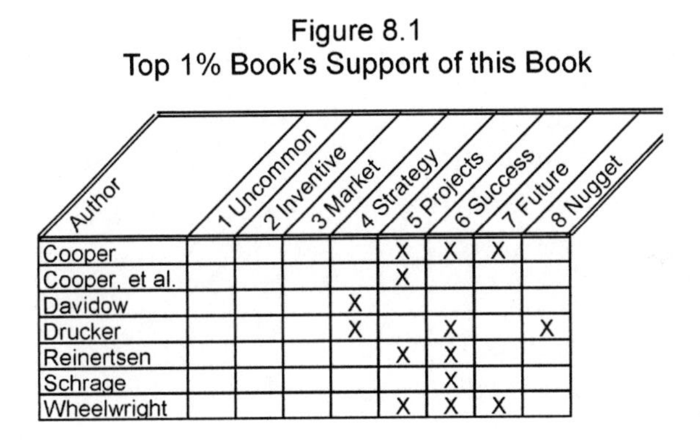

| Author | 1 Uncommon | 2 Inventive | 3 Market | 4 Strategy | 5 Projects | 6 Success | 7 Future | 8 Nugget |
|---|---|---|---|---|---|---|---|---|
| Cooper | | | | | X | X | X | |
| Cooper, et al. | | | | | X | | | |
| Davidow | | | | X | | | | |
| Drucker | | | | X | | X | | X |
| Reinertsen | | | | | X | X | | |
| Schrage | | | | | | X | | |
| Wheelwright | | | | | X | X | X | |

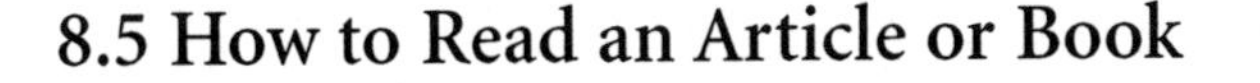

## 8.5 How to Read an Article or Book

Of course you know how to read! What I mean is how to read to make the best use of your time. And how to not clutter your mind with stuff that is not going to help you substantially improve your performance.

The great strength of these good books is that they are comprehensive. That is, they attempt to tell all there is to know. However, their great weakness is they get out of date. To stay up-to-date, an innovation manager needs to read new books and articles in leading journals. But there are too many for a busy manager to read them all. Here is my advice on how to decide what to read.

Earlier in this book, we showed that most inventors don't have many patents, most inventors don't produce at a high rate. It is also true that most articles don't have much new information and don't have much impact. Obviously, it's important for busy people to have some tools to decide which articles are ones that will be valuable for them to read. Here are some suggestions from my experience:

1. Read the introduction or first chapter, if there is no introduction. If it does not describe firm findings, don't read further. By firm findings, I mean specific results, preferably quantitative. At the very least, the author should claim to offer specific tools and techniques. Notice that the first chapter of this book starts with saying that the Inventor Profile describes 43% of the variance in inventors' productivity.
2. Read the conclusion. That's right; jump right to the conclusion. If it does not describe firm findings, don't read further. If both the introduction and conclusion describe firm findings, read the main part of the paper.
3. Examine the methodology. Many apparently quantitative results are meaningless because they are built on weak foundations. First, the study group may be irrelevant. Psychologists love to study their captive sophomore students. Do these young people have any relevance to inventive creativity? Unlikely. Second, the samples tend to be small. As we showed in this book, even a group of 100 R&D workers will contain only 36 inventors, among which only 4 will be prolific. Unless special measures are taken, the study will describe the least productive workers. Third, a notoriously weak technique is the blanket mail survey. Mail surveys are inexpensive, but are notoriously unreliable.[89] How does the researcher know that the respondent is qualified?

   Furthermore, the academic community is apparently unaware of one of the dark secrets in the private sector: high-level executives never answer surveys; in fact, they rarely even see them. The executive's administrative assistant either throws surveys away immediately or routes them to a member of the executive's staff. Most of these staff people will throw it away. In the rare event that someone actually follows up, they can always claim that it got lost. If they must do the work, it gets the minimal attention that it deserves, because it cannot improve their performance rating. In fact, responding to a survey has only downside risk; if proprietary or inaccurate information is revealed, the staff person can be criticized.

4. Look for results that matter. Finally, we quote Benjamin Disraeli, "There are three kinds of lies: lies, damn lies, and statistics." If an article is going to help you in your job, its results must be ones that make a measurable difference in your performance in a brief time. Thus, the quantitative results should explain a substantial proportion of the variation, like the 43% for the Inventor Profile. Unfortunately, the common t-statistic reported in regression calculations is much weaker; it measures the probability that there is a non-zero relationship between variables. When a result is reported as "significant at the 0.05 level" it means that the probability of there really being no relationship is 0.05, but the explained variance might be only 1%! In that case, 99% of the results are from other factors!
5. Reject most opinions. These tools for reading books and articles can be criticized because they reject all articles that are "opinion pieces." Yes, that is what I recommend. If the article's content is mostly opinion and no data, how do you know that it can help you? A lot of information for improving innovation is given in quantitative studies. These are the areas that will have the most impact on your performance over the next few years. So, they deserve the most investment of your time and effort.
6. Accept a few opinions. A possible exception to this "opinion rule" is an article by a prolific inventor, experienced manager of innovation or eminent author. Unfortunately, these articles are rare; most opinion articles are written by lesser academic professors or consultants. Even those articles written by experienced practitioners should be read with a critical eye for those opinions that can be applied to your company and can improve its competitive advantage. That is, many such experiences and opinions are

applicable only to the company where they occurred and perhaps even only for that technology, market and time.

Having given this advice, I remind you that what you learn from reading these books and articles is just Common Knowledge. If it is news to you, its benefit is raising you from the Swamp of Ignorance and Inefficiency, as depicted in Figure 1.3. However, to gain a competitive advantage, you must create Uncommon Knowledge.

## 8.6 Summary

The nuggets of knowledge from this chapter are that you can be a better manager of innovation. Become a leader. Develop courage. Read the top 1% books. Read other articles and books with an eye toward what they can do to improve your performance significantly. These three components—leadership, courage, and knowledge—are like the legs on a stool: each dependent on the other and all necessary for success, as shown in Figure 8.2.

Figure 8.2 Managing Innovation

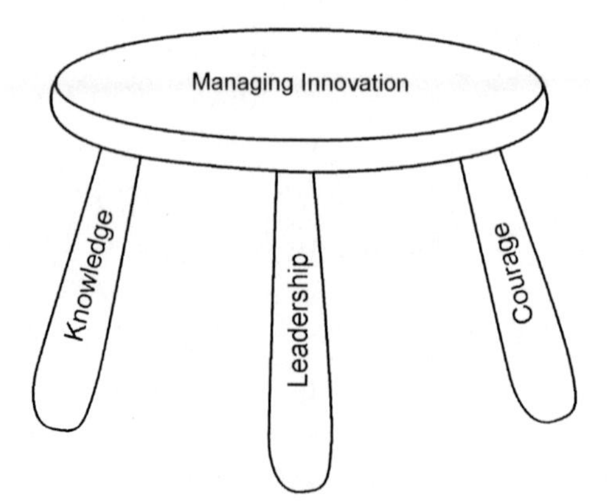

# 9

# *Mining for Companies That Manage Innovation Better*

In the previous chapters, the theme of this book has been how inventors and managers can improve how they manage their innovations. We have discussed the many ways that you can find better ideas, more ideas, and manage them into more successful products. Another source for finding better ways to manage innovation is to find which companies do it better and adopt their methods.

This sounds like a good idea, but it is difficult to put it into practice for several reasons. The best research so far indicates there has not been much improvement in new product success.[90] Also, there is not much difference between companies.[91] There are several reasons for this result. One, most new products are incremental improvements and few are the kind of innovations that we seek. Therefore, it is difficult to separate out the innovations. Two, the actual measurement of success in new products is also difficult. For example, how long after the sales begin do you decide if a new product is successful or not?

Another way to find companies that are good at managing innovation is to examine their patent productivity. Certainly, patents and innovations are not the same thing. Many patents describe inventive ideas that never become new products. And some new products are protected by trade secrets rather than patents. Nonetheless, the companies with strong innovation cultures tend to produce many patents.

When I founded the Institute for Invention and Innovation in 1996, one of our goals was to find ways to measure both the productivity of individual inventors and of companies. There were no good ways to do it at that time. However, we have developed several new techniques that are described in detail in Appendix C. In this chapter, we will discuss the highlights.

## 9.1 Patent Productivity of Inventors

Figure 9.1 Patent Productivity of Companies

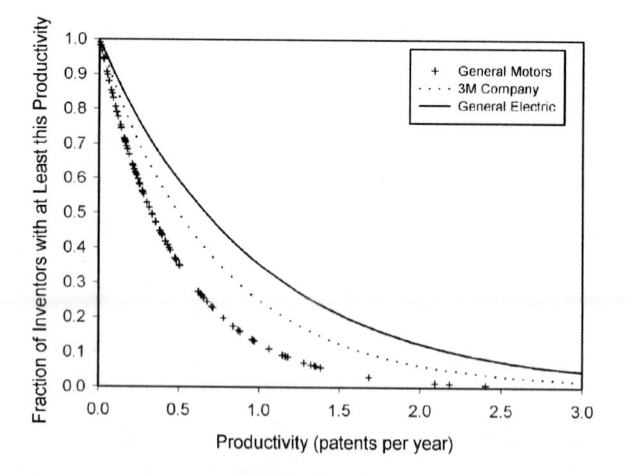


One of the most important facts about inventors is that their patent productivity has a wide range. We have examined the patent careers of over 20,000 inventors in over 30 companies. In every case, the distribution of patent productivity follows the exponential distribution, as shown in Figure 9.1 for three companies that cover the entire range of all those studied. This figure shows that most inventors produce at a very low rate and very few produce at a high rate. That is, the distribution of patent productivity is not at all like the common bell curve, as shown in Figure 9.2. Both these distributions have the same mean and standard deviation. Repeating from Chapter 3, only about 10% of inventors produce more than two patents per year. Yet, the top 10% inventors in every company produce over 50% of the patents. Thus, inventive talent is rare, but very productive.

Figure 9.2 Exponential Versus Normal Distribution

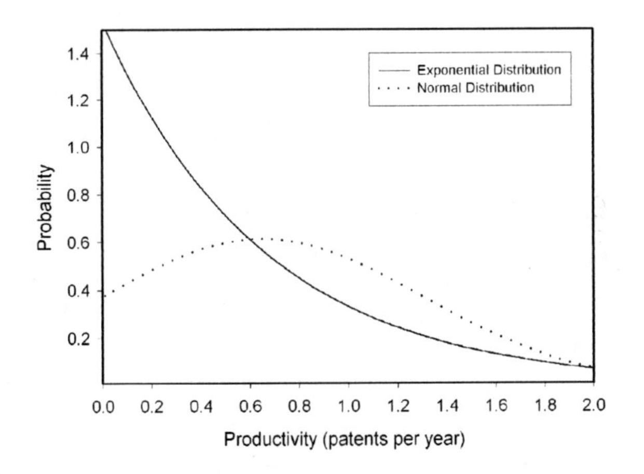


Another important fact about inventors is that the vast majority of them produce patents at the same rate over their entire careers. There is random fluctuation over time, but there is very little evidence of either learning or decline. Only 1.3% of the inventors had a measurable trend. And as many inventors had downward trends as upward.

Yet another important fact is that these results are also found for music composers and psychologists.[92] This is important, because in these professions there is little commercial opportunity, scarce funding, and almost no management. One would expect inventors to have a different result because they have the opposite conditions.

These are rather dismal findings. So we examined the patent productivity of these companies to try to find some differences.

## 9.2 Patent Productivity of Companies

There are three obvious conclusions from these results. One, no company is substantially better at not hiring low potential candidates. If it did, the left side of its curve in Figure 9.1 would bend down. Two, no company is substantially better at giving more resources to high productivity inventors. If it did, the right side of its curve in Figure 9.1 would bend up. All the companies we have studied differ only in their average patent productivity. Three, no company is substantially better at coaching or training its inventors. If it did, its inventors would have increasing patent productivity over time. No company exhibited a trend greater than 3%.

Certainly there are differences in the average patent productivity of companies. But the range is small. Of all the companies that we have studied, the most productive has only about twice the rate of the least productive. This is much smaller than the range of sales per employee, sales growth, profitability, return on stockholder's equity, and other measures of economic success. In fairness, most areas of management have had better tools than the R&D manager has had.

We have found some factors that have a substantial impact on patent productivity. One such factor is changes in strategy during which inventors must learn new technologies. This would cause slumps in an inventor's patent production over time. One of the companies we studied was Unisys, which has had many changes in strategy, as discussed in Appendix C. We detected substantial fluctuations in Unisys' time pattern of patent production.

This result tends to confirm that the methodology can detect the factors that have a substantial impact on patent productivity. Thus, our finding that there are no substantial differences tends to indicate that these companies do not manage innovation substantially differently from each other.

I don't want to appear to be critical of R&D managers. Indeed, I have over 25 years of experience in managing new product development. It is a difficult task and there have been few effective ways to measure performance. The methods described in this book may be a new aid to R&D management.

# 10

# *Conclusion: Mining the Nuggets*

Managing innovation is a lot like mining for gold. The bigger nuggets lie upstream, in Uncommon Knowledge. How do you create it? Remember the three components of successful innovation management: inventive idea, market need, conformance to a business unit's strategy. The three components do not flow in linear sequence, but rather in a circular, often messy, interchange.

1. Inventive Ideas.

- Find the most inventive people. Use the Inventor Profile as a tool. It explains 43% of the differences in performance. Chances are, the most prolific inventors will comprise about 3 percent of your lab.
- Check with inventors about their backlog of ideas. Ask for them informally in conversation—not in writing. Be positive about all voiced contributions.

2. Market Needs.
   - Don't try to copy your competitor's products, and don't analyze market conditions. These point to the past. Forget about traditional market research.
   - Live the life of a customer. What makes the customer frustrated? Talk to the people who train customers to use your products or to the sales rep who sells the product.
   - Use the Weak Link Law. Analyze existing products—yours and those of competitors—for fatal flaws.
   - Evaluate market problems by looking for the head-slap response.

3. Strategic Statement.
   - Does your business unit's strategic statement talk about where you want to be in the next five years? If not, write one that does.
   - Make sure it has wide support across functions—lab, marketing, sales, and manufacturing.

During the Fuzzy Front End, use a selection method such as Cooper's NewProd to assess an idea's uncertainties and its contribution to Uncommon Knowledge. Make sure the method is understood and accepted by everyone in your company. Then recruit the best people—usually the idea champions. For the rest of the team, select people with the necessary technical skills, but make sure they all get along with each other. Invest more resources in projects championed by highly productive inventors, and less in those of the less productive.

Once a project is selected, use a Stage-Gate system to guide the project through development. Give the project team freedom to run the show. Monitor progress with frequent, informal conversations. Expect setbacks and obstacles. Remember: The purpose of the early stages is to reduce uncertainty. Be sure the team addresses the right questions, rather than getting the right answers. Design experiments for a reasonable chance of *failure* (not to be confused with *mistakes*).

Make prototypes early and often. Show them to a marketer. Evaluate by looking for the head-slap response. When the marketer is satisfied, show the prototype to an insightful customer. First, however, obtain a non-disclosure agreement.

In the middle and later stages, use a project scheduling chart. Have the accounting department evaluate the project based on variations of net-present-value calculations.

For long-term improvement, invest wisely in product improvement and support. Don't let teams get bogged down in customer training or endless improvements. Look critically at existing products. Do they have substantial defects? Is there a logical next generation for some products? Don't be fooled. Customer requests for product improvements rarely amount to substantial sales.

Evaluate lab people. Consider transferring the less productive to other positions in the company. In hiring replacements, use the Inventor Profile. Hire people with expertise at the periphery of the company's key technologies to increase the in-house knowledge pool. Remember: Only people make innovations. If your company's culture says, "Innovation is expected here," you are more likely to generate plentiful and successful innovations.

Managing innovation is fundamentally a creative, dynamic process. You cannot manage it by command-and-control. Instead, be a coach. Size up individual players and invest in the best. Create a game plan. Stick to it even while others in the company emphasize your setbacks. Remind colleagues that the goal is Uncommon Knowledge—a competitive advantage that will earn more profits.

Finally, make yourself a better contributor. Learn to be a better leader. Have courage. Study the best books. Read new books and articles selectively.

# 11

## *Take Action—Start Mining*

What did you like about this book? What did you not like? Give me your feedback at jchuber@InventionAndInnovation.org.

Keep up to date on the latest news about inventors, invention and innovation. Visit our website at InventionAndInnovation.org.

If you gained new insights from this book, that's great. But just one person cannot change an entire company. One person alone can't create a good strategic statement. One person alone can't create a good project selection system. One person alone can't create a good Stage-Gate system. However, you can start the change. You will need to get your colleagues to learn what you have learned. Copy parts of the book and give them to your co-workers. Recommend the book to others. Buy an extra copy, highlight the key topics and drop it on the desk of someone who really needs to become a better nugget.

You'll benefit from taking this leadership role. The more people in your company who understand the better ways to manage innovation and mining for nuggets, the more successful you'll be. You'll have more ideas and better ideas. You'll have more interesting projects. You'll create Uncommon Knowledge. You'll increase your competitive advantage. And you'll have more successful new products. Best of all, it will be fun!

# *About the Author*

John C. Huber is Executive Director of the Institute for Invention and Innovation, which he founded in 1995. Its mission is to create the tools needed to effectively manage inventive and innovative performance. It provides these tools to the public through journal articles and books. Eleven articles and this book have been published so far. The latest research results are posted at InventionAndInnovation.org.

Today, such management practice is mostly based on each individual's experience. The quantitative information that does exist are not powerful predictors. The Institute seeks to find tools that are objective, quantitative, powerful, robust, and broadly-applicable.

Previous to founding the Institute, Dr. Huber had 30 years of experience in new product development at 3M Company. There he held technical, marketing or business management positions in nine different markets and thirteen different technologies. These ranged from fiber optic technology in the telecommunications market to flourochemical technology in the fabric protection market to laser holography technology in the blood analysis market. His principal achievement has been an outstanding record of new product introductions, with 79% successful out of 29 versus commonly-cited references reporting 20% successful. While at 3M, he authored one book and 36 articles, principally in the field of fiber optic telecommunications. Two of the articles received best paper awards.

Dr. Huber received BSEE, MSEE and PhD degrees from the University of Missouri—Columbia and an MBA degree from the

University of St. Thomas. He is a recipient of the Missouri Honor Awards for Distinguished Service in Engineering. He is listed in the 2001 edition of Marquis *Who's Who in the World* and the 2002 edition of Marquis *Who's Who in America.*

Dr. Huber can be contacted at jchuber@InventionAndInnovation.org.

# Appendix A

## *The Top 10% Books*

The top 1% of books are listed and described in Chapter 8. Here is a larger listing of the top 10% books that I've liked, but without comment on their particular strengths. Some of them may not be in stock at major bookstores, but you can find them in major libraries. They are separated by general category: Innovation, New Product Development, and Leadership.

### A.1 Innovation


Amabile, Theresa M. 1983. *The social psychology of creativity.* New York: Springer-Verlag.

Austin, James H. 1978. *Chase, chance and creativity: The lucky art of novelty.* New York: Columbia University Press.

Blohowiak, Donald W. 1992. *Mavericks! How to lead your staff to think like Einstein, create like da Vinci, and invent like Edison.* New York: Business One Irwin.

Christensen, Clayton M. 1997. *The innovator's dilemma: When new technologies cause great firms to fail.* New York: Harvard Business School Press.

Couger, J. Daniel. 1995. *Creative problem solving and opportunity finding.* New York: Boyd & Fraser.

Csikszentmihalyi, Mihaly, Rathunde, Kevin,& Whalen, Samuel. 1993. *Talented teenagers: The roots of success and failure.* New York: Cambridge University Press.

Drucker, Peter F. 1985. *Innovation and entrepreneurship.* New York: Harper & Row.

Feldman, David Henry, Csikszentmihalyi, Mihaly, & Gardner, Howard. 1994. *Changing the world: A framework for the study of creativity.* New York: Praeger.

Florman, Samuel C. 1994. *The existential pleasures of engineering.* New York: St. Martins Griffin.

Gamache, R. Donald, & Kuhn, Robert Lawrence. 1989. *The creativity infusion: How managers can start and sustain creativity and innovation.* New York: Harper & Row.

Gardner, Howard. 1993. *Frames of mind: The theory of multiple intelligences.* New York: Basic Books.

Gilman, John J. 1992. *Inventivity: The art and science of research management.* New York: Van Nostrand Reinhold.

Gretz, Karl F., & Drozdeck, Steven R. 1992. *Empowering innovative people.* New York: Probus Publishing.

Herzberg, Frederick. 1966. *Work and the nature of man.* New York: The World Publishing Co.

Josephson, Matthew. 1992. *Edison: A biography.* New York: John Wiley & Sons.

Kash, Don E. 1989. *Perpetual innovation: The world of competition.* New York: Basic Books.

Koestler, Arthur. 1964. *The act of creation.* New York: Dell Publishing Co.

Kuhn, Robert Lawrence, (Ed.). 1988. *Handbook for creative and innovative managers.* New York: McGraw-Hill.

Langer, Ellen J. 1997. *The power of mindful learning.* New York: Addison-Wesley.

Leonard-Barton, Dorothy. 1995. *Wellsprings of knowledge: Building and sustaining the sources of innovation.* New York: Harvard Business School Press.

Maslow, Abraham H. 1970. *Motivation and personality (2nd ed).* New York: Harper & Row.

Maslow, Abraham H. 1971. *The farther reaches of human nature.* New York: The Viking Press.

May, Rollo. 1975. *The courage to create.* New York: WW Norton.

McClelland, David C. 1961. *The achieving society.* New York: D Van Nostrand.

Morgan, Michael. 1993. *Creating workforce innovation.* New York: Business & Professional Publishing.

Newell, Allen, & Simon, Herbert A. 1972. *Human problem solving.* New York: Prentice-Hall.

Nonaka, Ikujiro, & Takeuchi, Hirotaka. 1995. *The knowledge-creating company: How Japanese companies create the dynamics of innovation.* New York: Oxford University Press.

Ochse, R. 1990. *Before the gates of excellence: The determinants of creative genius.* New York: Cambridge University Press.

Patterson, Marvin L. with Sam Lightman. 1993. *Accelerating innovation: Improving the process of product development.* New York: Van Nostrand Reinhold.

Perkins, D. N. 1981. *The mind's best work.* New York: Harvard University Press, Cambridge MA.

Plsek, Paul E. 1997. *Creativity, innovation, and quality.* New York: ASQC Quality Press.

Polya, G. 1957. *How to solve it: A new aspect of mathematical method, 2nd ed.* New York: Princeton University Press.

Price, Derek J. De Solla. 1963. *Little science, big science.* New York: Columbia University Press.

Price, Derek J. de Solla. 1986. *Little science, big science and beyond.* New York: Columbia University Press.

Rabinow, Jacob. 1990. *Inventing for fun and profit.* New York: San Francisco Press.

Rogers, Everett M. 1983. *Diffusion of innovations.* New York: The Free Press.

Root-Bernstein, Robert Scott. 1989. *Discovering.* New York: Harvard University Press, Cambridge MA.

Schank, Roger. 1988. *The creative attitude.* New York: Macmillan.

Schmoolker, Jacob. 1966. *Invention and economic growth.* New York: Harvard University Press, Cambridge MA.

Schon, Donald A. 1967. *Technology and change: The new heraclitus.* New York: Delacorte Press.

Shore, Sidney X. 1999. *Invent! Constructive discontent.* New York: Crisp Publications.

Simonton, Dean Keith. 1984. *Genius, creativity, and leadership: Historiometric inquiries.* New York: Harvard University Press.

Simonton, Dean Keith. 1999. *Origins of genius: Darwinian perspectives on creativity.* New York: Oxford University Press.

Simonton, Dean Keith. 1988. *Scientific genius: A psychology of science.* New York: Cambridge University Press.

Stalk, George Jr., & Hout, Thomas M. 1990. *Competing against time: How time-based competition is reshaping global markets.* New York: The Free Press.

Sternberg, Robert J. 1999. *Handbook of creativity.* New York: Cambridge University Press.

Sternberg, Robert J. 1996. *Successful intelligence: How practical and creative intelligence determine success in life.* New York: Simon & Schuster.

Sternberg, Robert J. (Ed.). 1988. *The nature of creativity.* New York: Cambridge University Press.

Torrance, E. Paul. 1976. *Guiding creative talent.* New York: Robert E. Krieger Publishing.

Tushman, Michael L., & Moore, William L. (Eds.). 1988. *Readings in the management of innovation.* New York: Ballinger Publishing.

VanGundy, Arthur B. 1987. *Creative problem solving: A guide for trainers and management.* New York: Quorum Books.

VanGundy, Arthur B. 1984. *Managing group creativity: A modular approach to problem solving.* New York: American Management Assn.

VanGundy, Arthur B. 1988. *Techniques of structured problem solving, 2nd ed.* New York: Van Nostrand Reinhold.

von Hippel, Eric. 1988. *The sources of innovation.* New York: Oxford University Press.

Waitley, Denis. 1995. *Empires of the mind: Lessons to lead and succeed in a knowledge-based world.* New York: William Morrow & Co.

Westcott, Malcolm R. 1968. *Toward a psychology of intuition: A historical, theoretical, and empirical inquiry.* New York: Holt, Reinhart, & Winston.

## A.2 New Product Development

Cooper, Robert G. 1993. *Winning at new products (2nd ed).* New York: Perseus Books.

Cooper, Robert G., Edgett, Scott J., & Kleinschmidt, Elko J. 1998. *Portfolio management for new products.* New York: Perseus Books.

Davidow, William H. 1986. *Marketing high technology: An insider's view.* New York: The Free Press.

Dean, James W. Jr. 1987. *Deciding to innovate: How firms justify advanced technology.* New York: Ballinger Publishing.

Gee, Edwin A., & Tyler, Chaplin. 1976. *Managing innovation.* New York: John Wiley & Sons.

Hall, John A. 1991. *Bringing new products to market: The art and science of creating winners.* New York: American Management Assn.

Peters, Thomas J., & Waterman, Robert H. Jr. 1982. *In search of excellence.* New York: Harper & Row.

Rastogi, P. N. 1995. *Management of technology and innovation: Competing through technological excellence.* New York: Sage Publications.

Reinertsen, Donald G. 1997. *Managing the design factory: A product developer's toolkit.* New York: The Free Press.

Rivette, Kevin G. & Kline, David. 2000. *Rembrandts in the attic: Unlocking the hidden value of patents.* New York: Harvard Business School Press.

Schrage, Michael. 2000. *Serious play: How the world's best companies simulate to innovate.* New York: Harvard Business School Press.

Tingstad, James E. 1991. *How to manage the R&D staff.* New York: AMACOM.

Wheelwright, Steven C., & Clark, Kim B. 1992. *Revolutionizing product development: Quantum leaps in speed, efficiency, and quality.* New York: The Free Press.

## A.3 Leadership

Barr, Lee and Norma. 1994. *Leadership development: Maturity and power.* New York: Eakin Press.

Bennis, Warren. 1989. *On becoming a leader.* New York: Addison-Wesley.

Bennis, Warren and Nanus, Burt. 1985. *Leaders: The strategies for taking charge.* New York: Harper & Row.

Bennis, Warren, & Biederman, Patricia Ward. 1997. *Organizing genius: The secrets of creative collaboration.* New York: Addison-Wesley.

Bennis, Warren, & Goldsmith, Joan. 1994. *Learning to lead: A workbook on becoming a leader.* New York: Addison-Wesley.

Covey, Stephen R. 1991. *Principle-centered leadership.* New York: Summit Books.

De Pree, Max. 1989. *Leadership is an art.* New York: Doubleday.

Drucker, Peter F. 1992. *Managing for the future.* New York: Truman Talley Books/Dutton.

Frangos, Stephen J. with Bennett, Steven J. 1993. *Team zebra.* New York: Oliver Wight Publications.

Kouzes, James M., & Posner, Barry J. 1987. *The leadership challenge.* New York: Jossey-Bass.

Oakley, Ed, & Krug, Doug. 1991. *Enlightened leadership: Getting to the heart of change.* New York: Simon & Schuster.

Pinchott, Gifford III. 1985. *Intrapreneuring.* New York: Harper & Row.

# Appendix B

## *The Inventor Profile*

This Appendix is reprinted with permission from a previously published article. [93]

### B.1 Introduction

One of the principal challenges to an R&D manager is hiring candidates with high potential and then managing them so their potential is realized. This challenge has a long history going back to the first corporate R&D labs.[94] Previously, some screening tools have been published.[95] But these have had generally low levels of prediction.

In this article, I describe the Inventor Profile, an objective and quantitative screening tool that can assist R&D managers in both hiring new employees and in identifying existing employees to staff new programs. The Profile is a powerful predictor of inventive potential and explains 43% of the variance. This result is much better than earlier R&D screening tools that explain 1–30% of the variance, and these tools often used variables that were difficult to implement in practice. About half of an inventor's Inventive Productivity can be predicted when they are new graduates.

The Inventor Profile can be used in three ways. First, it can be used in its questionnaire form. Second, it can be used to screen resumes. That is, the more powerful predictors can help a manager to pick the candidates with more inventive potential. For example, an extremely high undergraduate GPA is not a predictor of high Inventive Productivity. Third, it also can be used as a guide during an informal interview. That is, the more powerful predictors should be phrased as questions to the applicant. For example, "When did you first know you would become an engineer?"

## B.2 How the Inventor Profile Was Developed

The Inventor Profile was developed in three stages. First, using the published descriptions of successful inventors and scientists[96] and my own 28 years of experience in managing inventors, I developed a survey questionnaire. Second, the questionnaire was reviewed by six successful inventors and managers of inventors and then tested on thirty prolific inventors. Their suggestions for improvement were incorporated. Third, the actual Inventive Productivity for each questionnaire recipient was determined from his or her patent record.

The Inventor Profile has two important differences from previous work. First, each inventor's actual Inventive Productivity is used. Previous work has shown that eminent inventors have substantially larger productivity than that of average inventors.[97] In addition, publication productivity has been shown to be a useful indicator of scientific prominence.[98] In contrast, many other surveys have used the opinions of peers or managers to indicate technical prominence, a less precise method. Second, the questions and responses have a disciplined structure. The respondents selected one of several sentences that best described a particular variable. In contrast, many other surveys have used scales in which responses might be "Strongly Agree" or "Happens Frequently." A problem with such terms is that various people use them differently. These various interpretations can obscure the results. The use of strong anchoring phrases has been successful in R&D project selection,[99] so I adopted it for this study.

The Inventor Profile has a strong quantitative and statistical foundation. The U. S. patents for inventors are publicly available on the Internet at www.uspto.gov and in many patent repositories throughout the country. At the Institute for Invention and Innovation, we have examined the patent records for over 20,000 inventors in eleven large firms. Most inventors have few patents and most of the remainder have very low rates of Inventive Productivity.[100] Using stratified sampling, we created a normally distributed random sample of five hundred inventors with at least five patents over at least four years, with at least 0.4 patents per year, and having the most recent patent in the last five years. Thus, the sample represents the top 2.5% of inventors (500/20,000). The respondents were from such well-known firms as 3M Company, AMP, General Motors, Goodyear Tire and Rubber, and Medtronic. We believe this sample of firms is representative because a recent quantitative study[101] found there is little difference in Inventive Productivity between firms. We confirmed that the 112 respondents were a representative sample by showing that their distribution of Inventive Productivity is normally distributed with the same shape as the entire sample of 500 inventors, and not skewed by geographic region, firm where they work, or field of technology. The only incentive offered to the recipients was a copy of the completed analysis for their own use.

Then we performed multiple regression to calculate the Inventor Profile's predictive power. That is, each inventor's Inventive Productivity (patents per year) is the independent variable and each inventor's questionnaire responses are the dependent variables. Multiple regression calculates the variable coefficients that produce the best fit of predicted Inventive Productivity to the actual Inventive Productivity. As stated above, the Inventor Profile has substantially better predictive power than previous studies.

We received 112 responses, for a response rate of 23%, when undelivered Profiles were subtracted. While this response rate is less than that for studies within a firm that are sponsored by an executive, it is larger than the typical response for a consumer survey. We estimate that this sample represents about 20% of the candidate population of inventors in large firms. Thus, samples that are more than five times larger are not possible.

## B.3 Questionnaire Form of the Inventor Profile

You can get a copy of the questionnaire form of the inventor profile by sending me an email at jchuber@InventionAndInnovation.org. It is ready to copy onto 8.5 by 14 inch paper and staple into a booklet format.

## B.4 Scoring the Inventor Profile

The Profile is designed for prolific inventors who are busy and devoted to their work. Thus, it is brief and can be completed in less than fifteen minutes. The scores and coefficients are shown next to the response box and description for each variable. A positive coefficient means that particular variable contributes to Inventive Productivity, and a negative coefficient detracts from it. Obviously, the scores and coefficients were not shown to the surveyed inventors. The value for each variable is calculated by multiplying the response score by the coefficient. For example, if an inventor participated in sports for eight hours a week, the variable "sports" would have a value of 8 x -0.039 =

-0.312. An inventor's overall score is calculated by multiplying his or her score for each variable by its coefficient, summing the results and adding a constant value of –2.252. This overall score is that inventor's predicted Inventive Productivity. Each variable name is shown in boldface type.

**An Inventor's Early Characteristics**

Many times personal characteristics may not be apparent to each individual person, but can be found from examining a large number of people. The following questions are intended to find such common characteristics.

Please enter the number of hours per week in the box that you spent during the school year (not summer vacation) for ages 14–22 in the activities listed.

| Description of the variable | Score | Coef-ficient |
|---|---|---|
| I participated in team and individual **sports** for at least two years for x hours/wk. | hrs/wk | -0.039 |
| I practiced and played non-sports **games** (e.g. chess, cards, magic) for at least two years for x hours/wk. | hrs/wk | -0.095 |
| I participated in **youth** groups (e.g. Boy Scouts) for at least two years for x hours/wk. | hrs/wk | 0.070 |
| I read **books** not required in schoolwork for at least two years for x hours/wk. | hrs/wk | -0.057 |
| I **socialized** with friends (including dating) for at least two years for x hours/wk. | hrs/wk | -0.018 |
| I enjoyed **entertainment** (e.g. TV, spectator sports) for x hours/wk. | hrs/wk | 0.030 |
| I **worked** for pay for at least two years for x hours/wk. | hrs/wk | 0.017 |
| I practiced and played a **musical** instrument for at least two years for x hours/wk. | hrs/wk | -0.085 |
| I held a **leadership** role in an organization (student government, honorary society, social fraternity, etc.) for at least two years for x hours/wk. | hrs/wk | -0.059 |

As an inventor, you know that an **invention** must satisfy three tests to qualify for a patent – new useful and unobvious. But there are lesser inventions that don't meet these criteria. Please check the ONE statement that best describes your inventions for ages 14–22.

| Description of the variable | Score | Coef-ficient |
|---|---|---|
| An invention was patented. | check= 5 | -0.052 |
| I had an invention that was used, but not patented. | check= 4 | -0.052 |
| My inventions were interesting and fun for me, but not actually used. | check= 3 | -0.052 |
| I enjoyed designing inventions, but did not actually make them. | check= 2 | -0.052 |
| My inventive interests appeared later in life. | check= 1 | -0.052 |

What was your scholastic performance in college? My undergraduate GPA was ____.

| Description of the variable | Score | Coefficient |
|---|---|---|
| **GPA** | Value of GPA | -0.728 |

Please check the ONE statement that describes your college **degree**(s) best.

| Description of the variable | Score | Coef-ficient |
|---|---|---|
| I earned a PhD degree and top scholastic honors (e.g. GPA greater than 3.8/4.0) | check= 7 | 0.203 |
| I earned an MS degree and top scholastic honors (e.g. GPA greater than 3.8/4.0) | check= 6 | 0.203 |
| I earned a BS degree and top scholastic honors (e.g. GPA greater than 3.8/4.0) | check= 5 | 0.203 |
| In addition to a degree marked above with top scholastic honors, I earned advanced degree(s), but not with top honors. | check= 4 | 0.203 |
| I earned top scholastic honors only in courses related to the field of my degree. | check= 3 | 0.203 |
| I earned a degree, but not with top honors. | check= 2 | 0.203 |
| I do not have a college degree. | check= 1 | 0.203 |

What degree of variety occurred in your college education? Please check ALL the statements that describe you.

| Description of the variable | Score | Coef-ficient |
|---|---|---|
| I knew what field I would study from the beginning and concentrated on it **exclusively**. | check= 1 | 0.195 |
| **I switched** majors at least once. | check= 1 | 0.195 |
| I took at least two **elective** courses that were unrelated to my major at the time. | check= 1 | 0.192 |
| I have degrees in two or more **majors**. | check= 1 | 0.411 |
| I attended two or more **colleges**. | check= 1 | -0.145 |
| The first college I attended was at least 100 **miles** from my home. | check= 1 | 0.246 |

What made you **choose** your **field**? Please check the ONE statement that is the best description of your decision.

| Description of the variable | Score | Coef-ficient |
|---|---|---|
| I liked the fact that my field helps me to discover mistakes and errors in existing methods. I like to correct and improve the ways problems are solved. | check= 1 | 0.145 |
| I liked the fact that my field helps me to discover new ways of doing things that have not been thought of before. An important part of this is developing a new method that replaces a clumsy method. | check= 2 | 0.145 |
| I liked the fact that my field helps me to discover new ways of doing things that have not been thought of before. An important part of this is the intellectual challenge of being the best. | check= 3 | 0.145 |

| Description of the variable | Score | Coefficient |
|---|---|---|
| My first patent issued when I was ________ years of **age**. | Value of age | -0.023 |

How important was **mentoring** for your first patent? Please check the ONE statement that best describes your first patent.

| Description of the variable | Score | Coef-ficient |
|---|---|---|
| Another inventor with many patents was a co-inventor and the team leader. | check= 4 | -0.033 |
| I received substantial guidance and advice from another inventor with many patents, but he/she was not a co-inventor. | check= 3 | -0.033 |
| Another inventor or manager assigned the task to me, but did not provide substantial guidance or advice. | check= 2 | -0.033 |
| I came up with the problem and solution entirely on my own. | check= 1 | -0.033 |

How was your first patent **related** to your prior education? Please check the ONE statement that is the best description.

| Description of the variable | Score | Coef-ficient |
|---|---|---|
| It was based on my college research and was filed while I was in college. | check= 4 | 0.291 |
| It built on my college research, but the invention occurred during my employment. | check= 3 | 0.291 |
| It depended on my major field, but was not built on actual college research. | check= 2 | 0.291 |
| It was outside my major field. | check= 1 | 0.291 |

What was the nature of **support** you had while you worked on your first patent? Please check the ONE statement that is the best description.

| Description of the variable | Score | Coef-ficient |
|---|---|---|
| My manager provided all the equipment, assistance, etc. without my asking. | check= 5 | 0.108 |
| My manager provided all the equipment, assistance, etc. but I had to request and justify it. | check= 4 | 0.108 |
| I received some support, but having more would have helped substantially. | check= 3 | 0.108 |
| I received no significant support, but was allowed to work on my own. | check= 2 | 0.108 |
| My manager was opposed to my work; I had to do it in my spare time. | check= 1 | 0.108 |

| Description of the variable | Score | Coefficient |
|---|---|---|
| I am left-**handed**. | check= 1 | 0.281 |
| I am right-**handed**. | check= 2 | 0.281 |
| I generally work x **hours/week**. | Value of hrs/wk | 0.049 |

One of the factors is technical **management**. Please check the ONE statement that best describes your most recent experience.

| Description of the variable | Score | Coef-ficient |
|---|---|---|
| My management usually supports invention and innovation. Project selection is done by a standardized process that I usually understand and agree with. I have enough support services to do my work effectively. | check = 4 | 0.182 |
| My management fluctuates in its support of invention and innovation. The project selection process seems to produce decisions that I sometimes don't understand. Sometimes support services for my work are withdrawn and disrupt my work. | check = 3 | 0.182 |
| My management only weakly supports invention and innovation. The project selection process is weak and produces decisions that I often don't understand or agree with. I have to fight and scrounge for support services. | check = 2 | 0.182 |
| My management's actions demonstrate no real support for invention and innovation. Any inventing that I accomplish is entirely on my own. | check = 1 | 0.182 |

Another factor is the role of **marketing** in finding good problems to work on. Please check the ONE statement that best describes your most recent experience.

| Description of the variable | Score | Coefficient |
|---|---|---|
| Our marketing/sales people clearly identify new product opportunities. The requirements are fairly stable during the product development process. We respect and trust one another. Our new product introduction process actually helps get the job done. Marketing/sales works hard to make the product introduction a success. | check = 4 | 0.105 |
| Our marketing/sales people identify opportunities, but they are often vague. We often have to start the design over because of changes. We are able to work together, but are wary of one another. Our new product introduction process seems to hinder as much as help. Marketing/sales doesn't seem to work hard at making the product introduction a success. | check = 3 | 0.105 |
| Our marketing/sales people mostly come up with weak ideas that are mostly copies of what the competitors already sell. It seems that they are never satisfied with our designs. We dislike each other. Our new product introduction process is a barrier to progress. Marketing/sales mostly ignores new products until someone gets a sale. | check = 2 | 0.105 |
| Our opportunities are driven by technological capabilities. Marketing has no significant role. Our inventions are driven through the organization by sheer will power. Marketing/sales seems to be usually indifferent to our inventions. | check = 1 | 0.105 |

Another factor is organizational direction and **stability**. Please check the ONE statement that best describes your most recent experience.

| Description of the variable | Score | Coef-ficient |
|---|---|---|
| Our business unit has a clear mission and strategy. I understand it and it helps direct my work. We have clear assignments and communications. | check = 4 | -0.217 |
| Our business unit has a vague mission and strategy. It does not substantially help direct my work. Our official assignments and communications are also vague. But with a few trusted colleagues, we have charted our own course and the business unit has agreed to it. | check = 3 | -0.217 |
| Our business unit has a vague mission and strategy. It does not substantially help direct my work. Our official assignments and communications are also vague. Some of us have tried to fill the gap, but were unsuccessful. | check = 2 | -0.217 |
| Our business unit changes vision and strategy frequently. I either frequently redirect my work or I ignore the changes. Our official assignments and communications also change frequently. I spend a lot of time educating new bosses and colleagues. | check = 1 | -0.217 |

What thing would MOST **improve** your Inventive Productivity? Please check ONE statement.

| Description of the variable | Score | Coef-ficient |
|---|---|---|
| Larger salary and/or a bonus for each patent. | check= 1 | 0.178 |
| Fewer meetings and fewer reorganizations. | check= 2 | 0.178 |
| Faster funding approval and/or a manager that champions programs. | check= 3 | 0.178 |
| More assistants and/or being the team leader. | check= 4 | 0.178 |

There is some controversy over the occurrence of **inspiration** or a sudden insight toward solving a problem. Please check the ONE statement that best describes your overall experience.

| Description of the variable | Score | Coef-ficient |
|---|---|---|
| I have never had an inspiration and I suspect that it is just a myth for others. Problems get solved by systematically applying known principles in a variety of combinations. | check = 1 | 0.141 |
| Once in a while a new way of solving a problem will come to me. But it is more like a vague hint than what most people think of as an inspiration. Still, most problems get solved by systematically applying known principles in a wide variety of combinations. | check = 2 | 0.141 |
| Ideas for solving problems occasionally pop into my head, often at odd hours or while doing something else. They are often merely clues, rather than complete solutions. | check = 3 | 0.141 |
| I actively cultivate inspirations. I use specific techniques, such as: deliberately putting the problem aside, meditation, going on a long drive. | check = 4 | 0.141 |

There is also some controversy over the effectiveness of structured group techniques of enhancing creativity, such as **brainstorming**. Please check the ONE statement that best describes your experience.

| Description of the variable | Score | Coef-ficient |
|---|---|---|
| I have participated in brainstorming or other structured creativity techniques. They frequently generate a breakthrough solution that had stumped the best people. | check = 4 | 0.022 |
| I have participated in brainstorming or other structured creativity techniques. Sometimes they generate a new idea. I question their efficiency. | check = 3 | 0.022 |
| I have participated in brainstorming or other structured creativity techniques. Mostly I find them a waste of time. Most of the ideas are well-known. | check = 2 | 0.022 |
| I have never participated in brainstorming or other structured creativity techniques. | check = 1 | 0.022 |

There is also some controversy over the effectiveness of structured individual techniques of **enhancing** creativity. Please check the ONE statement that best describes your experience.

| Description of the variable | Score | Coef-ficient |
|---|---|---|
| I regularly use "pretending I am the product" or other structured individual creativity techniques. They are important to nearly every significant advance. | check = 3 | 0.631 |
| I have tried "pretending I am the product" or other structured individual creativity techniques. Mostly there is a small fraction of good ideas. | check = 2 | 0.631 |
| I have never tried "pretending I am the product" or other structured creativity techniques. | check = 1 | 0.631 |

Some inventors prefer to focus closely on one problem. Others prefer to work on several **problems** at once. Please check the ONE statement that best describes your preference.

| Description of the variable | Score | Coef-ficient |
|---|---|---|
| I prefer to focus on one problem at a time. It takes a lot of effort to collect all the relevant data and to immerse myself in the problem. This focus allows me to organize the data and carefully plan my approach. | check = 1 | 0.060 |
| I prefer to work on two or three problems at once. When one problem hits a roadblock, I put it aside and focus on another. I have a system for keeping each problem organized. | check = 2 | 0.060 |
| I prefer to work on four or more problems at once. I find the variety of switching back and forth keeps me from becoming bored or burned out. Sometimes it gets chaotic, but I find it more stimulating than disruptive. | check = 3 | 0.060 |

Inventors also vary based on the way they collect data and start **designing** or experimenting. Please check the ONE statement that best describes your preference.

| Description of the variable | Score | Coef-ficient |
|---|---|---|
| First, I collect all the available data, including a thorough patent search. Then I organize the data to identify the key variables. Finally, I set up a set of designs or experiments to measure the variables. The results of these designs or experiments will identify the solution. | check = 1 | 0.278 |
| First, I collect just enough data to have a general knowledge of the problem. Then I try a variety of simple designs or experiments to get involved with the nature of the problem. Based on the success of these, I collect more data in those areas. Then the designs or experiments become more complex and lead to the solution. | check = 2 | 0.278 |
| First, I try to think of a new method that doesn't use the methods of the existing solutions. Then I usually have to discover the boundaries and limits of the new method by my own experiments. The comparative success of these experiments versus the existing solutions provides a guide for further work. | check = 3 | 0.278 |

Some inventors prefer working as part of a **team** of inventors. Others find it more productive to work alone. Please check the ONE statement that best describes your overall experience.

| Description of the variable | Score | Coef-ficient |
|---|---|---|
| I prefer to work alone on inventions. When I have to work with a team, I dislike the time and effort spent on setting direction and reporting results. I also am uncomfortable with styles of thinking that are very different from mine. Sometimes there is competition for credit. | check = 1 | -0.057 |
| Teams are useful for getting input, data, and specific skills from others. But the actual invention is still a solitary chore. It's best if one person is in charge of the actual design. | check = 2 | -0.057 |
| I prefer to work with a team. I find that others have knowledge or skills that I don't have and which are important to solving difficult problems. I like to bounce ideas off others and have a lively give-and-take discussion. It is important for colleagues who work together smoothly. | check = 3 | -0.057 |

## B.5 Scores and What They Mean

All the variables explain 43% of the variance in the actual rate of patent production, as shown in Figure B.1. We created a new variable, named **field focus,** as the sum of **choose field** and **majors.** It has a more significant impact than these two variables alone.

Figure B.1 Predicted versus Actual Patent Productivity

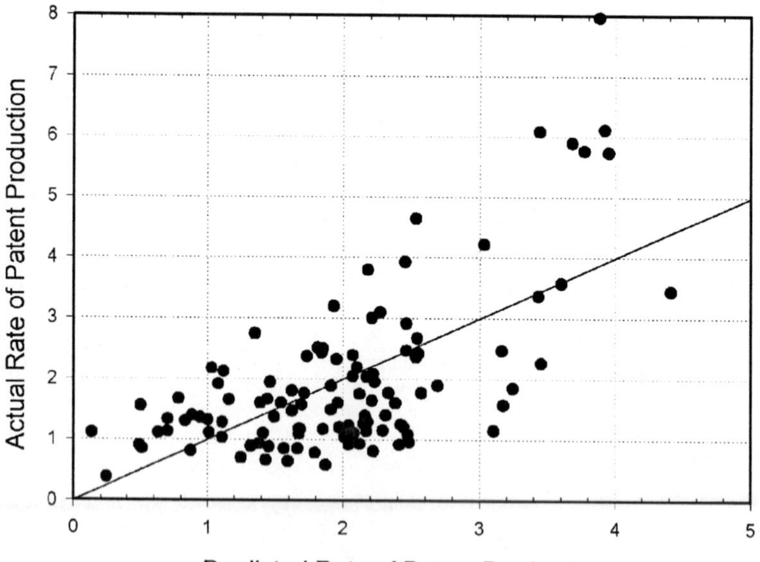


The principal findings from the Profile are summarized in Figure B.2. The twelve variables with the largest impact account for 75% of the explained variance. Having an advanced degree with honors predicts 25% of an inventor's patent productivity. That, combined with working many hours per week, predicts 37%. Each further variable contributes less and less. Figure B.3 shows these variables rearranged, with those appropriate for new hires to the left and those for experienced people to the right. About half of an inventor's patent productivity is predictable when they are new graduates and the other half when they have substantial experience.

Figure B.2 Effect of Variables in Order of Impact

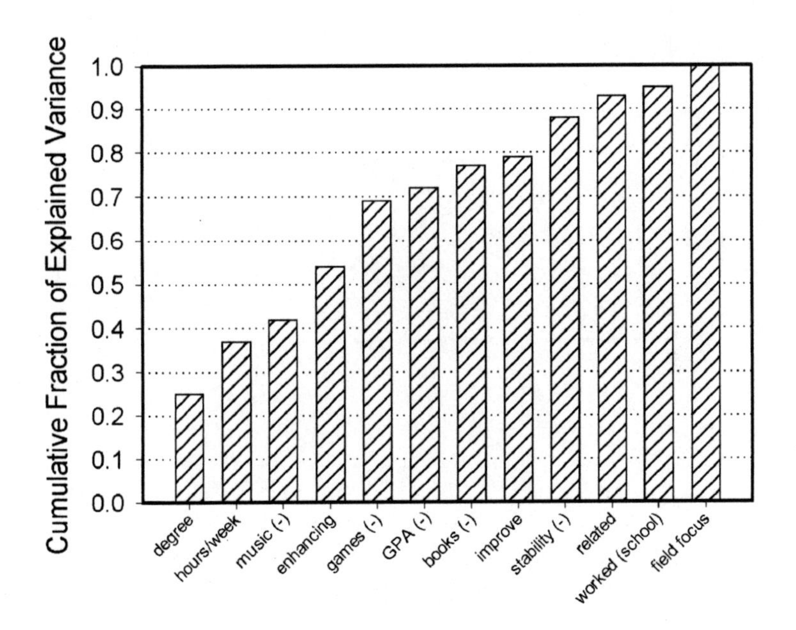

Figure B.3 Effect of Variables in Order of Experience

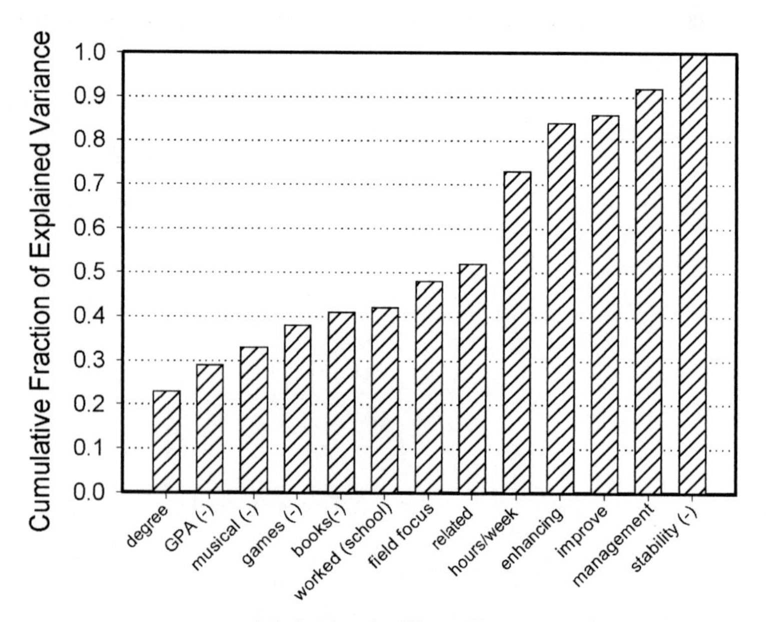


This study is based on the responses from 112 prolific inventors. They represent the top 2.5% of all inventors and so have already achieved eminence in their field.

These inventors' general characteristics are not surprising. They have a strong desire for excellence (61% chose their field to "be the best"). They are well-educated (45% have graduate degrees with honors; 90% have an undergraduate GPA of 3.0 or higher). In school, they spent little time away from their studies (average 5.6 hours per week). Hard work is common (average 50 hours per week). They work in a supportive environment (63% had strong management support on their first patent; 56% report strong, consistent management support of innovation). That is, they work in a culture that says "Innovation is expected here."[102] However, these inventor's assessments of marketing support and a stable strategic direction had a wide range. They are team players (56% preferred working on a team). When asked to identify the most important thing to improve their Inventive Productivity, 33% chose faster decisions or supportive management; only 9% chose getting bonuses or royalties; 28% chose having fewer meetings or reorganizations; 17% chose having more assistants or being the team leader.

However, there are some surprising results. Some of the common beliefs about inventors are either found to be non-predictive or even negative predictors. Perhaps most surprising, is that the highest undergraduate GPA is a predictor of generally lower Inventive Productivity, though 90% had a 3.0 or better and 45% had graduate degrees with honors. This result is consistent with a previous study of engineers.[103] A possible explanation is that inventor's inquiring minds and eagerness to experiment tend to interfere with their getting the highest undergraduate GPA, but fit well into the more investigative graduate program.[104] Less than 15% reported good results from cultivating inspirations, brainstorming, or visualization. Also, time spent playing a musical instrument, playing chess or bridge games, or reading books outside their coursework are predictors of generally lower Inventive Productivity. Another surprising result is that a very stable business strategy is a predictor of generally lower Inventive Productivity. A possible explanation is that a very long-persisting strategy may result in exhausting a firm's technological opportunities. Among the other common beliefs that are non-predictors of high Inventive Productivity are: time spent playing sports, time in youth groups, having the first patent at an early age, being left-handed, having a supportive relationship with marketing, and preferring to work on many problems at the same time.

Undoubtedly, there are many productive inventors who are musicians, bridge champions, chess masters, omnivorous readers, or Eagle Scouts. But this study indicates that these achievements are not good predictors of high Inventive Productivity in general.

One of the surprising results from the Inventor Profile is that many variables that were thought to be good predictors of Inventive Productivity are not. We performed a statistical analysis that examines these poor predictors. The variables are ranked in descending order of impact in Figure B.4.

Figure B.4 Poor Predictor Variables

| variable | t-stat | p-level | $r^2$ |
|---|---|---|---|
| problems | 1.69 | 0.09 | 0.03 |
| handed | 1.68 | 0.10 | 0.03 |
| socialize | -1.65 | 0.10 | 0.02 |
| choose field | 1.47 | 0.15 | 0.02 |
| miles | 1.41 | 0.16 | 0.02 |
| inspiration | 1.15 | 0.25 | 0.01 |
| marketing | -0.95 | 0.34 | 0.01 |
| age | -0.87 | 0.39 | 0.01 |
| brainstorming | -0.86 | 0.39 | 0.01 |
| designing | 0.75 | 0.45 | 0.01 |
| majors | 0.72 | 0.48 | 0.01 |
| team | 0.60 | 0.55 | 0.00 |
| org. leader | 0.57 | 0.57 | 0.00 |
| youth | 0.53 | 0.60 | 0.00 |
| support | 0.50 | 0.62 | 0.00 |
| inventing | -0.47 | 0.64 | 0.00 |
| mentoring | 0.35 | 0.73 | 0.00 |
| entertainment | -0.22 | 0.83 | 0.00 |
| sports | -0.18 | 0.85 | 0.00 |
| elective | -0.04 | 0.97 | 0.00 |
| colleges | 0.04 | 0.97 | 0.00 |

The term, p-level, is the probability that there is no relationship to actual Inventive Productivity. The term, $r^2$, is the fraction of explained variance. For example, let's examine the variable **problems.** Working on more than one problem at a time is like having many irons in the fire. It does contribute to higher Inventive Productivity, but only explains 2.5% of the variance. Thus, it has a negligible practical significance.

* * *

Any managers who want to get a copy of the Inventor Profile for their own use may send an email to me. They will receive the Profile as a Microsoft Word™ 6.0 document and the scoring as a Microsoft Excel™ 5.0 workbook. An important feature of this form of the Profile is that it includes four places for open-ended responses. When I developed the Profile, I found that inventors would participate best when given these opportunities for individualized responses.

* * *

## B.6 Conclusions

These 112 inventors had a wide range of Inventive Productivity, from 0.4 to 8.0 patents per year. The Inventor Profile can predict 43% of this variance in Inventive Productivity. This is better than earlier R&D screening tools, which explain 1–30% of the variance. An important improvement is that these variables can be easily evaluated, versus some earlier studies that used vaguely defined factors such as "commitment to work." Furthermore, twelve of the variables account for 75% of the explained variance. Strong positive predictors are: graduate degree with honors, work many hours per week, use of visualization to enhance their creativity, desire for independence, having their patent work related to their education, working their way through school, and field focus (80% either knew their field from an early age or switched majors into it). The Profile has several places for open-ended comments. The only new variable was the need for persistence. As one inventor put it, "Never, never, never give up!"

The Inventor Profile's strong predicting power arises from three factors. First, the measure of an inventor's performance is Inventive Productivity (patents per year), versus total patents which mixes productivity with career duration. Also, productivity is an objective fact, not a subjective assessment. Second, stratified sampling allows the analysis to focus on the most productive inventors. Third, strong anchoring phrases reduce the ambiguity in responses.

The high impact variables can be arranged in the time sequence that they can be evaluated. Thus, the variables that can be evaluated at hiring account for about half the explained variance in an inventor's Inventive Productivity. The rest of the explained variance arises from variables that can be evaluated once an inventor's career is well advanced. Equally important, the predictive variables are straightforward and easy to evaluate, even in an informal interview.

The unexplained variance is consistent with the wide variety of paths to inventive success. Another contribution to the unexplained variance is that much of inventing arises from tacit knowledge, that which we know but cannot clearly explain or describe.[105] Language-based questionnaires like the Inventor Profile are unlikely to discover much about tacit knowledge.

An important issue is patent quality versus quantity. Of course, patent quality is very difficult to evaluate from public data. But the cost of filing and maintaining a patent is substantial and mitigates against producing low-value patents. The average cost of filing a U. S. patent is $ 7,000 (www.bpmlegal.com). However, multinational companies need to protect their inventions around the world; 3M Company reports an average cost of $140,000 per patent, not including any litigation.[106] Furthermore, the inventors in the top 10% of productivity produce over 50% of the patents for the firms in this study.

Another important finding of this study is the impact of the highly skewed distribution of total patents and rate of patents, discussed earlier. That is, most inventors have few patents and most produce at a very low rate. These skewed distributions mean that a simple survey will describe the least productive people. Since the least productive people are clustered closely together and are such a large proportion of any simple sample, a simple analysis will largely ignore the highly productive people who are un-clustered and rare. Without stratified sampling, none of the Inventor Profile variables are significantly predictive.

We have shown that the 112 respondents to this study correspond to much larger samples of inventors in general. However, larger studies are planned. We are especially interested in examining how the Profile fits the inventors in a single firm and how quality may be directly addressed.

# Appendix C

## *Patent Productivity of Individuals and Companies*

### C. 1 Abstract

We have developed new and improved statistical methods to make more accurate measurements of patent productivity and more powerful comparisons between companies.[107] We chose several companies as examples of a range of inventive cultures. The distribution of inventor's patent productivities (patents per year) across a company is exponential, and not the normal (bell curve) distribution. The range of the company's overall patent productivity (patents per year per inventor) was only 2:1. We show that quality and productivity are related. Each individual inventor's patent production is generally constant over his or her career, random over time, with a time pattern that generally follows the Poisson distribution. Less than 2.5% of any company's inventors exhibited a measurable trend, and that overall trend itself was less than 3%. There was little evidence of management investing extra resources in high productivity inventors. The most significant effect was for companies with substantial changes in strategy, which exhibited substantial deviation from the Poisson distribution, and which may be due to fluctuations in inventors' time pattern of productivity.

## C.2 Introduction

Studying the productivity of inventors has a long history,[108] but there have been few comparisons between organizations and the differences were not subjected to statistical tests. Also, the motivations and methods of inventors and other R&D workers have a long history,[109] there have been few comparisons between organizations and the differences were not subjected to statistical tests. Furthermore, the reports of methods for managing inventors and R&D workers tend to report on one company's experience or one person's opinion,[110] with no comparison between companies. Indeed, measuring R&D effectiveness itself has been difficult.[111] In addition, studying the successes and failures of new product development (an outcome of many inventions) has a long history,[112] however actual improvement has come into doubt.[113]

There are two main causes for this lack of comparison. For the case of successes and failures of new product development, much of the data is gathered from mail surveys, which have the weakness of unknown quality of the respondent. Market researchers have struggled with this problem without significant success.[114] Even if the respondent is perceptive and truthful, he or she may not have knowledge that is both broad and detailed. For example, a typical new product development project may take three years and have a low probability of success, meaning that respondents with fifteen years of experience may have personal knowledge of only one successful project.

For the case of inventive productivity, patents have been shown to be a useful indicator of a company's success.[115] However, a mitigating factor is that not all patents are related to successful products.[116] Although patent data is factual and a matter of public record, the statistical methods have been weak. We use several new and improved methods that make comparisons that are statistically powerful.

We show that inventive productivity (patents per year) is not distributed evenly, but more surprisingly, does not follow the normal (bell curve) distribution. For every one of the 32 companies we have studied, inventive productivity follows the exponential distribution. That is, most inventors have very low productivity and very few have high productivity. The difference between companies in overall patent productivity can be tested and is statistically significant. However, the range among companies that are perceived to have a strong inventive culture and those that are not is only about 2:1.

In all discussions of personal productivity, the issue of quality versus quantity arises. There is an enduring specter of a hack cranking out huge volumes of low quality work. We show that inventive quality and quantity are closely related. The inventors in the top 10% of productivity produce about 50% of a company's patents. Since each patent's cost is in the range of $7,000 to $140,000, good management practice would indicate these are good investments. Furthermore, eminent inventors have productivity about twice that of average inventors.

We also measure several other parameters of inventors, including career duration, randomness of outputs, and fit to the Poisson distribution. But perhaps the most significant finding is that less than 2.5% of inventors exhibit any measurable trend in their patent output, and that overall trend is less than 3%. These findings tend to indicate that companies' policies and cultures have little effect on inventors' patent productivity. It also tends to indicate that individual talent is the most significant factor in determining an inventor's personal productivity and the company's collective productivity. The results from this study show that we cannot reject the allegation that inventions are just "harvested from the wild" as opposed to being effectively managed.[117]

## C.3 Choosing the Samples

We have found that a company must have at least about 1,000 inventors for effective analysis. This translates into companies with at least $2 Billion in annual sales. In addition, most patents are produced by manufacturing companies. So our population of companies is limited to the manufacturers in the Fortune 500. We selected seven companies, each one as an example of an inventive culture or experience.

1. 3M Company (Minnesota Mining and Manufacturing) is well known for its innovative climate.[118] In addition, it has had only one substantial restructuring, the 1995 spin-off of Imation. Both companies' patents are included as 3M.
2. AMP (now a part of Tyco International) is well known for its innovation in electrical connectors, especially in the computer and telecommunication fields. It experienced substantial restructuring in the mid-1990s, but the effect on patents is delayed about five years, the time for projects to be completed and any patent to be issued. So the effect may be small. Only electrical connector patents were included in the sample.
3. Goodyear was selected because it scored high marks on CHI Research's Tech-Line™ 1998 Academic Sampler (www.chiresearch.com), a measure of inventive performance.[119]
4. General Motors was selected as a counter-example because it has been widely criticized for continuing management methods that may not be best practices.[120] The company has many far-flung locations. We believe that the locations nearest corporate

headquarters would be most consistent with corporate policies. So, only inventors located in Michigan were included in the sample.

5. Unisys was selected as a counter-example because it is an extreme example of a company that has been merged, restructured, endured substantial layoffs, and had several dramatic changes in strategic direction.[121] Following the merger of Univac and Burroughs, the non-computer businesses of Unisys were sold off. Only the patent classes that endured after the 1989 merger were included in the sample.
6. General Electric was selected because it has elements of all the companies above. It had the very first industrial R&D lab.[122] However, beginning in the mid-1980s, some major businesses were sold and new ones added.[123] In addition, it has many far-flung locations. We believe that the locations nearest corporate headquarters would be most consistent with corporate policies. So, only inventors located in New York were included in the sample. However, these inventors contributed over one-third of GE patents.[124]
7. Petroleum was selected because of three reasons. First, it claims to have the highest patent efficiency in its industry (www.phillips.com). Second, one of its inventors was awarded the National Medal of Technology (www.ta.doc.gov) and has published his views on selecting inventive people.[125] Third, its R&D function suffered downsizing in the mid-1980s, but without substantial changes in strategic direction.[126] Thus, Phillips may be a contrast to Unisys and General Electric.
8. We generated a sample of eminent inventors who are recipients of the National Medal of Technology (www.ta.doc.gov), Inventors Hall of Fame (www.invent.org), Lemelson-MIT Prize (www.mit.edu), Industrial Research Institute Achievement Award

(www.iriinc.com), or Intellectual Property Owners Award (www.ipo.org).

Each sample's patents for the period 1976-1999 were found in the U. S. Patent Office database (www.uspto.org). These companies are specific examples chosen from our database of over thirty U. S. companies and over 20,000 inventors. However, the findings described below are typical. No company has been discovered that is an outlier from the range of parameters of these companies.

## C.4 Inventive Productivity

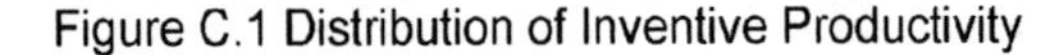
Figure C.1 Distribution of Inventive Productivity

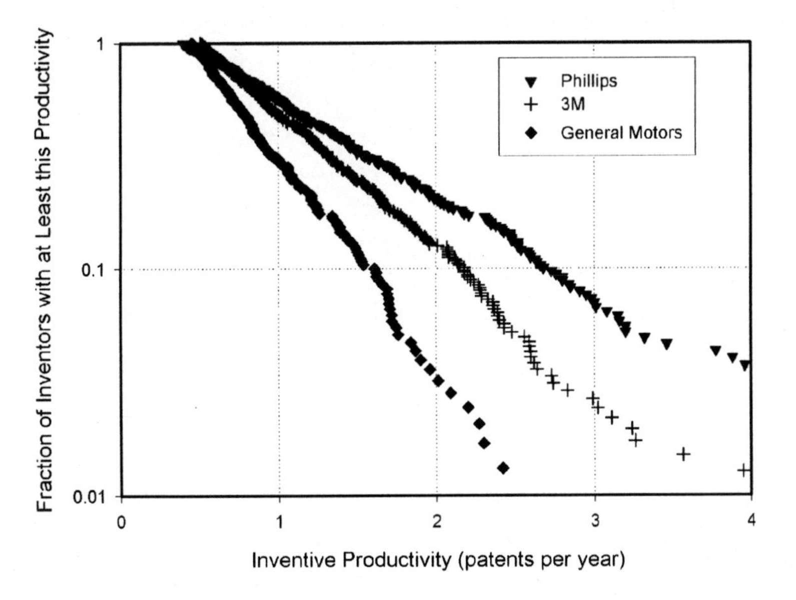


Obviously, the productivity for an inventor with only one patent cannot be calculated. We have found that it takes about five patents

spread over at least three years to get a reasonably accurate estimate for an inventor's productivity. Figure C.1 shows the cumulative distribution of inventive productivity for Phillips, 3M and General Motors. The straight-line form on semi-log scales indicates a good fit to the exponential distribution. We made a least-squares best fit to the most linear region and measured the goodness-of-fit for the entire distribution. The methods of the statistical treatment are described in the Statistical Details section. Table C.1 shows the parameter of the exponential distribution (phi = average patents/year) and the goodness-of-fit (g-o-f) for the samples. The parameter phi is consistent with the average patents per inventor used by Porter and Stern.[127] Larger values of goodness-of-fit indicate a better fit. For example, 3M's value of 0.95 means that the probability is 0.95 that the deviation from the straight line is due to just chance fluctuation alone. Thus, small values would indicate that a systematic process caused the observed deviation. An example of a systematic process would be extra investment in high productivity inventors. These are excellent goodness-of-fit values and indicate that systematic effects are small.

Table C.1 Parameters for Inventive Productivity

| Company | # inventors | phi | g-o-f |
|---|---|---|---|
| 3M | 432 | 0.72 | 0.95 |
| AMP | 199 | 0.58 | 0.50 |
| Eminent | 72 | 1.60 | 0.90 |
| General Electric | 629 | 0.96 | 0.50 |
| General Motors | 265 | 0.48 | 0.75 |
| Goodyear Tire | 132 | 0.62 | 0.95 |
| Phillips Petroleum | 334 | 1.04 | 0.50 |
| Unisys | 207 | 0.71 | 0.25 |

A natural question is whether the differences in phi are significant. Table C.2 shows the results of the statistical test on the exponential parameter.[128] The companies are ranked by increasing phi, with company name, number of inventors, and phi in the first three rows. Each cell in the table is defined by the companies for its row and column. The cell value is the probability that the difference between the companies' phi values is due to just chance fluctuation alone. Thus, small values in the table indicate significant differences in phi. For example, eminent inventors are different from every company so much that the odds are over 100:1 that it is just a lucky draw for the eminent inventors. Similarly, the odds that General Motors (with the smallest phi) is really the same as AMP (with the next smallest phi) are 50:1.

Table C.2 Comparison of Inventive Productivity Across Companies

| Company | General Motors | AMP | Goodyear Tire | Unisys | 3M | General Electric | Phillips Petroleum | Eminent |
|---|---|---|---|---|---|---|---|---|
| # inventors | 265 | 199 | 132 | 207 | 432 | 629 | 334 | 72 |
| phi | 0.48 | 0.58 | 0.62 | 0.71 | 0.72 | 0.96 | 1.04 | 1.60 |
| General M | | 0.02 | 0.01 | 0.00 | 0.00 | 0.00 | 0.00 | 0.00 |
| AMP | 0.02 | | 0.27 | 0.02 | 0.01 | 0.00 | 0.00 | 0.00 |
| Goodyear | 0.01 | 0.27 | | 0.11 | 0.07 | 0.00 | 0.00 | 0.00 |
| Unisys | 0.00 | 0.02 | 0.11 | | 0.44 | 0.00 | 0.00 | 0.00 |
| 3M | 0.00 | 0.01 | 0.07 | 0.44 | | 0.00 | 0.00 | 0.00 |
| General E | 0.00 | 0.00 | 0.00 | 0.00 | 0.00 | | 0.12 | 0.00 |
| Phillips | 0.00 | 0.00 | 0.00 | 0.00 | 0.00 | 0.12 | | 0.00 |
| Eminent | 0.00 | 0.00 | 0.00 | 0.00 | 0.00 | 0.00 | 0.00 | |

However, these differences need to be understood in context. It is well known that companies have substantially different cultures promoting invention. But, it is also true that companies have different standards for what inventions should be patented. A company may have a culture promoting invention, but also have a policy that only inventions with large business impact are patented. Another factor is that some companies tend to protect some inventions, not with patents, but as trade secrets, which makes their patent production less than their invention production. These factors can affect one another such that a company that strongly promotes invention may have a phi value nearly the same as another company that only moderately promotes invention.

What is more important is that all the companies shown here and over 25 others all follow an exponential distribution of inventive productivity. The importance of finding that the exponential distribution is a good fit is dramatized in Figure C.2, where the common probability distributions for the exponential and normal (bell curve) distributions are shown with the same parameters (mean and standard deviation of

0.65). For the normal distribution, most of the inventors are near the mean and few are much smaller or much larger than the mean. However, for the exponential distribution, most of the inventors are at small values, fewer are at the mean and still fewer are larger than the mean.

Figure C.2 Patent Productivity

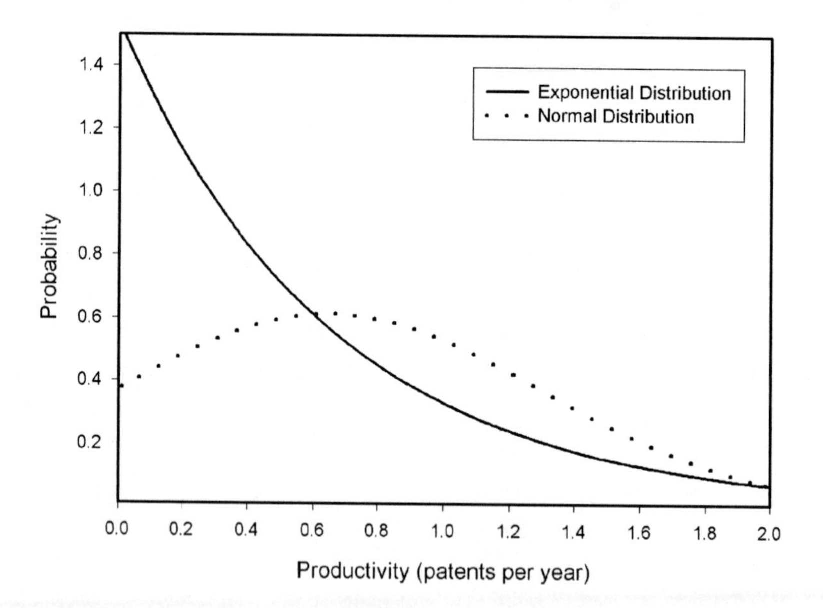


What the distributions in Figure C.1 show is that companies are not very good at investing additional resources in high productivity inventors to increase their productivity further. If effective investment did occur, the distribution would curve upward at high productivity. Of the more than thirty companies that we have studied, only Phillips exhibited a substantial additional investment in high productivity inventors, and that was for the upper 4% on the plot, who are the top 1% of all Phillips inventors. There is no doubt that companies have the intention, expectation, and effort toward such investment. However, the patent data shows significant results for only one company out of more than thirty.

Up to this point, we have focused on productivity and thus on the quantity of patents. In all discussions of human performance, there is the specter of a hack pumping out high volumes of low quality work. In the next section, we show that quality and quantity are closely related.

## C.5 Quality and Quantity of Inventive Productivity

As can be seen from Table C.1, the average phi for the companies is 0.73 and the sample of eminent inventors has phi of 1.60, or 2.2 times the average company. This ratio is the same as Nobel laureate scientists' scientific papers (both before and after their award) compared to average scientists.[129]

We also examined the number of patents produced by the inventors in the top 10% of productivity (patents per year). The proportion of each company's patents that are produced by the top 10% inventors is shown in Table C.3.

Table C.3 Proportion of Company's Patents from Top 10% Inventors

| Company | %Patents from top 10% inventors |
|---|---|
| 3M | 64.2 |
| AMP | 57.8 |
| General Electric | 58.5 |
| General Motors | 50.8 |
| Goodyear Tire | 57.4 |
| Phillips Petroleum | 55.2 |
| Unisys | 49.2 |

The average company has 57% of its total patents produced by the inventors in the top 10% of productivity. A typical patent costs between $7,000 and $140,000, including filing and maintenance fees.[130] If high productivity inventors tend to produce low quality, then the company's management is making poor investments.

Obviously, there are some inventors with low productivity whose inventions have created industries. Also, there are some inventors with high productivity whose inventions have had insignificant impact. However, these findings tend to support a positive relationship between high productivity and high quality. A similar relationship has been found for general creativity.[131] This relationship is not hard and fast, but it is the way to bet.

## C.6 Inventive Longevity

In the past, little attention has been paid to an inventor's career longevity. It was generally accepted that high productivity and a long career result in many patents, but there was little quantitative analysis.

Two important issues arise in the analysis of career duration. The first issue is the fact that some inventors may begin their careers or end their careers outside the limits of the 1976-1999 time interval. The fields of statistics named Survivor Analysis and Reliability have been created to treat this problem.[132] The second issue is to estimate the effort before the first patent and after the last. If an inventor's estimated start date and estimated end date are within the 1976-1999 interval, then that inventor's complete career has been observed.

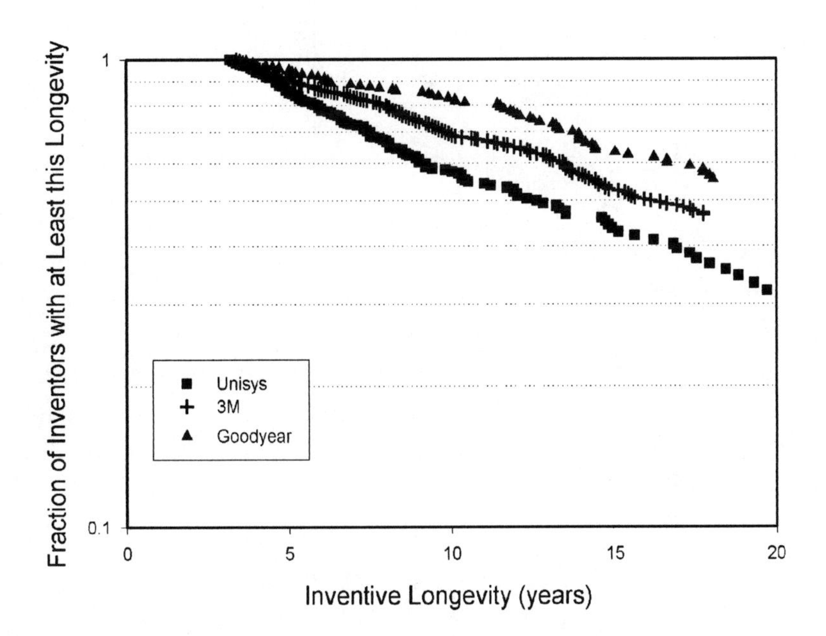


Figure C.3 Distribution of Inventive Longevity

The methods to resolve these issues have been reported elsewhere[133] and a brief description is in the Statistical Details section. Figure C.3 shows the cumulative distribution of inventive longevity for Unisys, 3M and Goodyear. The straight-line form on semi-log scales indicates a good fit to the exponential distribution. We made a least-squares best fit to the most linear region and measured the goodness-of-fit for the entire distribution. Details of the statistical treatment are described in the Statistical Details section. Table C.4 shows the parameter of the exponential distribution (tau = years) and the goodness-of-fit (g-o-f) for the samples. Larger values of goodness-of-fit indicate a better fit, as discussed in the previous section on productivity.

Table C.4 Inventive Longevity Across Companies

| Company | # inventors | tau | g-o-f |
|---|---|---|---|
| 3M | 502 | 19.2 | 0.53 |
| AMP | 199 | 13.4 | 0.70 |
| Eminent | 109 | 42.0 | 0.50 |
| General Electric | 732 | 17.5 | 0.01 |
| General Motors | 338 | 16.1 | 0.10 |
| Goodyear Tire | 191 | 33.3 | 0.01 |
| Phillips Petroleum | 275 | 15.4 | 0.56 |
| Unisys | 143 | 11.5 | 0.01 |

Those companies with poor fits (e.g. 0.01) exhibit two linear segments, indicating a mixture of two populations. The steeper segment is formed by inventors with shorter longevity and the less-steep segment is formed by inventors with longer longevity. Two of the companies with poor fits (General Electric and Unisys) had substantial restructuring with some inventor's having shorter careers than those remaining throughout the restructuring.

These exponential distributions show that most inventors have short careers and very few have long careers. It is important to emphasize that the term longevity used here describes these inventors only while they are inventing. A substantial portion of the work of a manufacturing company's R&D is not expected to result in a patent, such as product improvements. Also, some inventors start inventing later in their careers. Some inventors move to management, marketing, or manufacturing. Thus, a given inventor's overall career would generally be longer than the longevity estimated here.

## C.7 Randomness and the Time Pattern of Patents

One of the enduring issues in personal performance is the question of learning, decline, or stability.[134] Put in statistical terms, the question is whether the time pattern of each inventor's patents is random or not.

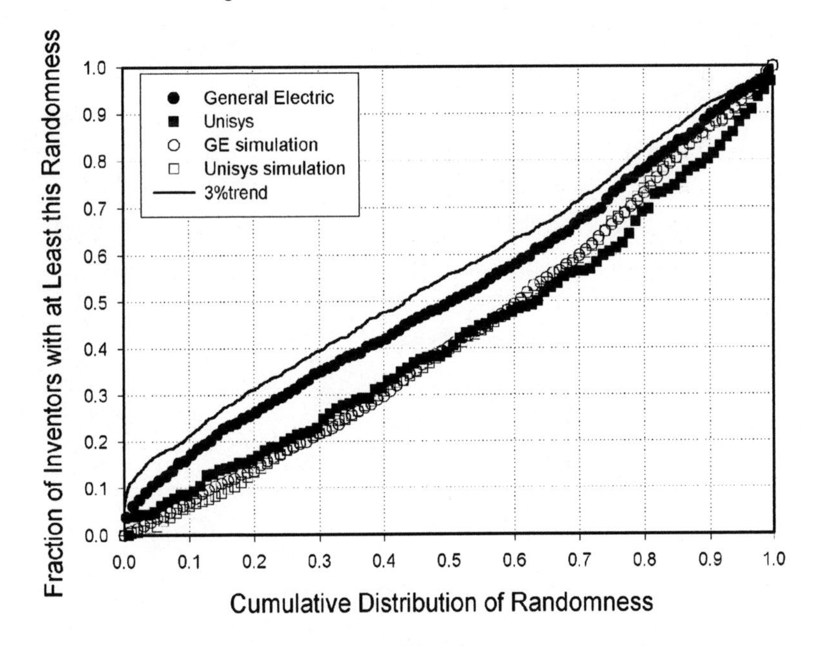


Previously, the most powerful methods for measuring randomness applied only to continuous variables, but patent counts are a discrete variable. Recently, an improved method of measuring the randomness of discrete variables has been discovered.[135] Using this method, we calculated the distribution of randomness for each companies' inventors. Figure C.4 shows the distribution for Unisys and General Electric. For clarity, the distributions are plotted at increments of 0.01. For comparison, we performed simulations—inventive productivity was drawn from an exponential distribution and inventive longevity was drawn from an exponential distribution; each company's phi and tau parameters were those calculated in the previous sections. Further details are given in the Statistical Details section. A company with a purely random pattern amongst its inventors will have a distribution the same as this simulation. General Electric exhibits a marked step in the lower tail; Unisys has a small step that is nearly negligible. Using other simulations, we have found that the inventors in this extreme lower tail exhibit trends in their output. For comparison, a 3% trend is shown. As we shall show below, these are a small fraction of the inventors. The other companies exhibited similar results.

Figure C.5 Distribution of Poisson-ness

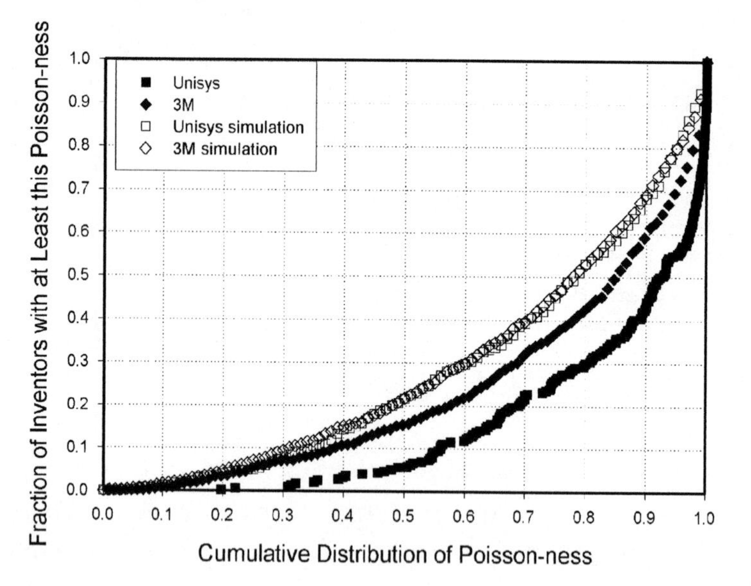


Another method for detecting systematic processes is to examine how closely the time pattern of patents follows the Poisson distribution.[136] The Poisson distribution is commonly found to describe similar discrete processes, such as telephone calls, accidents and nuclear radiation. Figure C.5 shows the distribution for 3M, Unisys and their simulations. For clarity, the distributions are plotted at increments of 0.01. A company with a purely Poisson pattern amongst its inventors will have a distribution the same as its simulation. General Electric exhibits a large step in the upper tail; 3M has a smaller deviation from its simulation. Simulations have shown that the inventors at the extreme upper tail exhibit excess fluctuations in their output. As we shall show below, this deviation is small for many companies. The other companies in this study and others in our database exhibited similar results.

Since both the distributions of randomness and Poisson-ness detect non-randomness in an inventor's time pattern, we can combine them for a more powerful selection process. After examining many companies, we have found that the following criteria are useful to select the "random" inventors: Poisson-ness less than 0.999 and ratio of Poisson-ness to randomness less than 40. Using these criteria, the distributions of randomness are shown in Figure C.6. All companies exhibited a similar improvement in randomness. For comparison, the "random" inventors from the previous 3% trend simulation are shown also. Using simulations, we have discovered that trends smaller than 2% do not distort these distributions significantly. The deviation from the ideal case shown in Figure C.6 is expected: the 3% trend has less impact for inventors with short careers and/or low inventive productivity and these criteria are not able to select them efficiently.

Figure C.6 Distribution of Randomness for Random Inventors

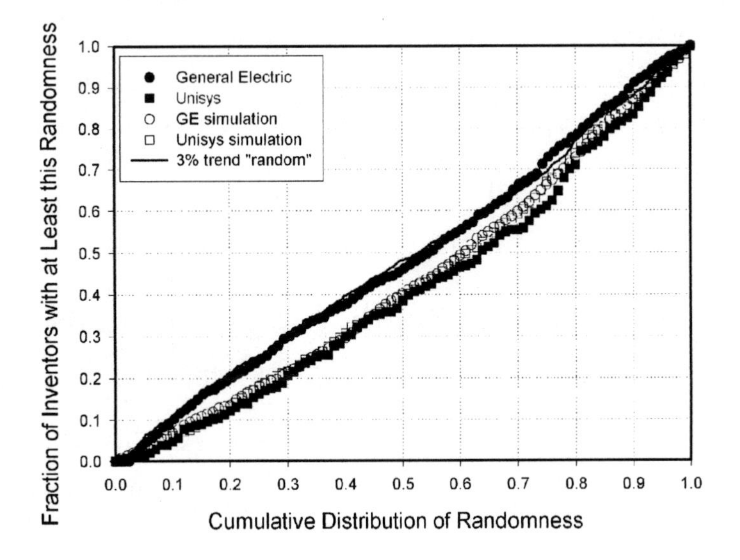

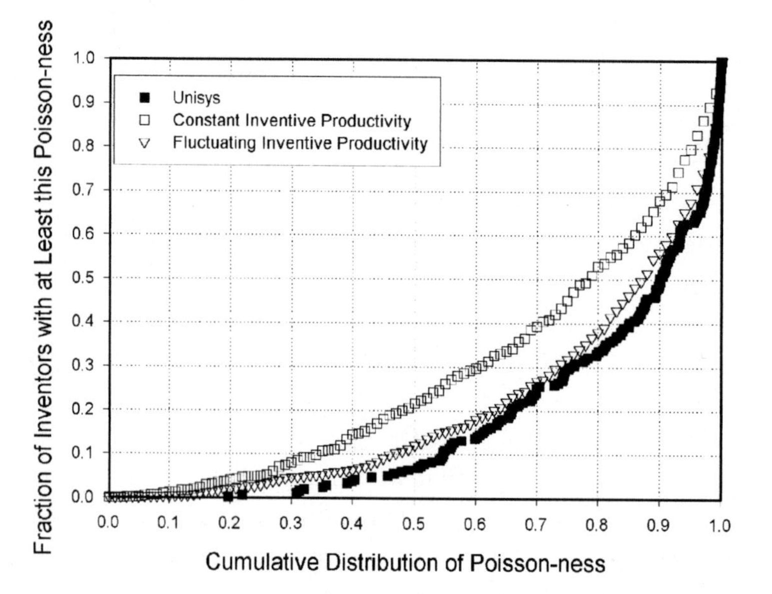


But for Poisson-ness, the situation is more complicated. For some of the companies (3M, AMP, Goodyear and Phillips) the distribution of Poisson-ness also improved. But for others (General Electric and Unisys), a substantial deviation remained. One likely cause of excess fluctuations is changes in strategic direction that force inventors to switch to new areas (e.g. new technologies), require new learning, and impose slumps on their time pattern of patents. Figure C.7 shows the distribution of Poisson-ness for Unisys, a simulation with a constant inventive productivity for each inventors, and a simulation with a fluctuating inventive productivity. The description of this simulation is in the Statistical Details section. General Electric had a similarly improved fit to a simulation with fluctuating inventive productivity. These two companies were specifically chosen as examples which had endured substantial restructuring and reorganization and thus could be expected to exhibit this fluctuation. Table C.5 shows the goodness-of-fit (g-o-f) for each company's "random" inventors to a constant-productivity simulation based on that company's parameters phi and tau. The g-o-f value is the root-mean-square-deviation. That is, 3M's inventors overall deviations from ideal randomness and Poisson-ness is 0.04. The other entries are discussed below.

Table C.5 Summary of Randomness, Poisson-ness and Trend

| Company | # inventors | g-o-f Random | g-o-f Poisson | # non-random | % non-random | % with trend |
|---|---|---|---|---|---|---|
| 3M | 7793 | 0.04 | 0.04 | 77 | 1.0 | 0.4 |
| AMP | 1,049 | 0.05 | 0.06 | 29 | 2.8 | 1.4 |
| General Electric | 2,830 | 0.06 | 0.10 | 153 | 5.4 | 2.1 |
| General Motors | 3,352 | 0.02 | 0.08 | 58 | 1.7 | 0.3 |
| Goodyear Tire | 1,146 | 0.04 | 0.07 | 27 | 2.4 | 1.3 |
| Phillips Petroleum | 1,307 | 0.05 | 0.08 | 62 | 4.7 | 2.5 |
| Unisys | 2,309 | 0.03 | 0.14 | 40 | 1.7 | 0.8 |

Examining Table C.5, we notice that the companies that were chosen for strong inventive cultures (3M, AMP, Goodyear), have the smaller goodness-of-fit errors. Similarly, the companies chosen for substantial changes in strategy (Unisys, General Electric) have large errors in their goodness-of-fit to the Poisson distribution. We can put these g-o-f values in context by comparing them to simulations. That is, the 3% trend simulation with non-random "inventors" removed exhibited residual deviation from the non-trend simulation. Also, a simulation with excess fluctuation had residual deviation from the constant productivity simulation. The average root-mean-square-error for randomness is 0.04. Thus, g-o-f values above 0.04 are likely to be caused by "random" inventors with small, but non-zero, deviations from the pure random case. The average root-mean-square-error for Poisson-ness is 0.08. Thus, g-o-f values above 0.08 are likely to be caused by "random" inventors with excess fluctuations from the pure Poisson case. Thus, General Electric stands out for trends amongst its inventors and Unisys stands out for excess fluctuation.

It is important to emphasize that both the individual inventor and the company have strong motivations to break away from a simple random time pattern of patents. They would hope to improve performance with experience, thereby exhibiting a trend. They would hope to have inventions with outstanding potential such that they had associated patents to support it or to block competition,[137] thereby exhibiting excess fluctuation. These are the inventors with non-random patterns and we now focus our attention on them.

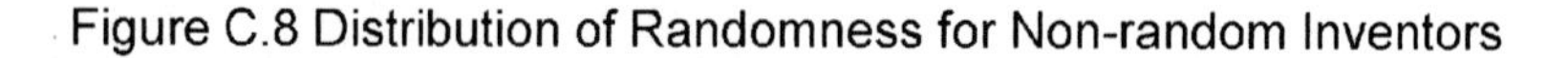
Figure C.8 Distribution of Randomness for Non-random Inventors

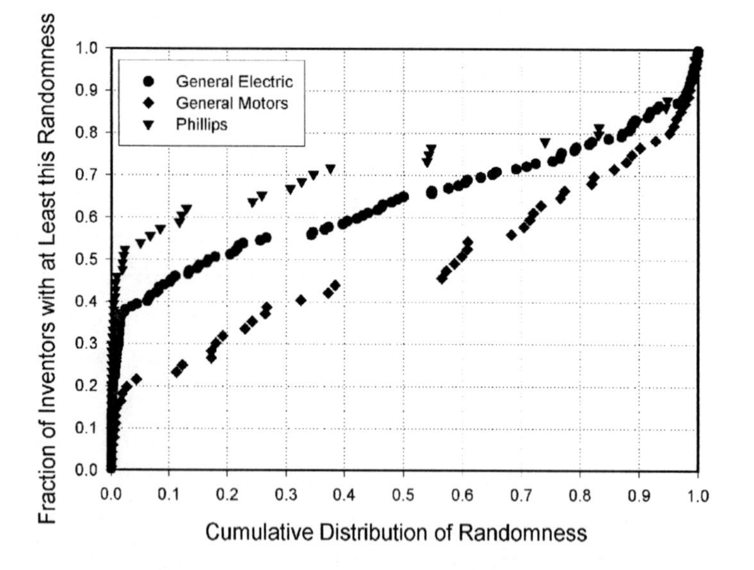


Figure C.8 shows the distribution of randomness for the non-random inventors at General Electric, General Motors and Phillips. That is, these inventors had Poisson-ness greater than 0.999 or ratio greater than 40. There are three distinct regions in the plot. Those inventors in the left tail exhibit a distinct trend. Those inventors in the right tail exhibit a single spike with much smaller patent production in other time elements. Those inventors in the middle exhibit strong fluctuations in their inventive productivity; probably they were the ones most affected by the changes in strategy.

Examining Table C.5 shows that the percentage of each company's inventors that are non-random is less than 3% except for General Electric and Phillips. Furthermore, the percentage of each company's inventors that exhibit a non-random trend is less than 2% except for General Electric and Phillips. For all companies, the non-random inventors with trends average 1.3% of inventors. The maximum overall trend is less than 3%, estimated from comparing General Electric to its simulation in Figures C.4 and C.6. Thus, there is little evidence of learning or decline in inventors.

The preceding sections used statistical tools to show that individual inventor's patent production fits standard distributions, including: conformance to random production, Poisson distribution over time, exponential distribution for productivity and longevity. However, the results are still just an empirical description. It might be an accident or an artifact. In the next two sections, we proceed to show that the findings are not really surprising and indeed there is a theoretical basis for these results.

## C.8 Randomness and Individual Effort

For those who labor long and hard on inventive work and for those who manage inventors, the notion of random variables may be offensive. The previous statistical analysis does not mean that inventive achievements "just happen out of the blue." It may appear that way from time to time. But some individuals have much larger productivity than others. And some individuals enjoy a much longer career than others. And some individuals exhibit trends or excess fluctuations. Except for companies with dramatic changes in strategy, these individuals occur no more often than one would expect in a large group of inventors, as discussed in the previous section.

We have no quantitative or objective way of measuring the inventive qualities of individual inventors. We can only measure their outputs. What the preceding statistical analysis implies is that these qualities exist and vary from individual to individual. That is, randomness and statistical distributions are the same methods that we also use to measure the distributions of IQ, height, weight, etc. across individuals. These are convenient models and do not imply that personal improvement is impossible. In fact, personal effort and other such systematic process can be detected by deviations from the standard statistical distributions.

The finding that inventive productivity is constant also needs explanation. These samples represent individuals who are at the highest levels in their fields. Such individuals are extremely gifted, tend to learn the teachable skills quickly, and thereafter are substantially self-taught. Thus, many inventors will exhibit a rate of increase or decrease that is smaller than these statistical methods can detect.

Then what explanation is there for these excellent fits to standard statistical distributions? One plausible answer arises from extreme value theory and the statistics of exceedances, as is discussed in the next section.

## C.9 Theoretical Support for the Exponential Distribution

We propose that there are two general variables that are important for a successful inventive career, loosely termed talent and tenacity. There have been many lists of the specific talents, abilities, or skills that make up these general variables.[138] We further propose that the general variable talent will be manifest by productivity (patents per year). Similarly, the general variable tenacity will be manifest by longevity. We propose that however the specific skills combine into talent or tenacity (e.g. multiplicative or additive, correlated or not, etc.), the result is a continuous distribution. We further propose that only those people in the general population with the highest levels of the important skills will exceed our criteria of at least five patents spread over at least three years. Thus, our analysis here only describes the upper tails of these distributions.

The upper tails of all common "textbook" distributions are described by the Generalized Pareto Distribution, of which the exponential distribution is an important special case.[139] This field is known as the statistics of exceedances.[140] Inventors with at least five patents are very rare, being less than 0.5% of the general population.[141] Thus, regardless of the shape of the distribution of talent and tenacity in the general population (at least as they apply to invention), it is not surprising that the upper tail follows the exponential distribution. The exponential distribution also describes productivity and longevity in scientists, composers and athletes.[142] Thus, there is support for the statistics of exceedances describing many fields of human performance.

## C.10 Conclusions

This study has used new statistical methods to examine the patent production of inventors and draw conclusions about how effectively they are managed. Our fundamental proposition is that if inventors are effectively managed, there will be significant differences between companies that are perceived to have different innovative cultures. However, the overall finding is that companies with strong innovative cultures had much the same patent production as companies with weak innovative cultures. We proceed to discuss the findings in detail.

For all the companies, the distribution of inventive productivity (patents per year) is exponential, not the normal (bell curve) distribution.

This means that most inventors produce patents at a very low rate, few produce at the average rate and very few produce at twice the average. This is a very different situation from the common experience with the normal distribution. That is, the most commonly encountered (i.e. "typical") inventor will have an inventive productivity well below the average, and generally will not exhibit the characteristics that lead to high productivity.

Furthermore, these companies had a goodness-of-fit to the exponential distribution that is exceptional. No company, even ones that are perceived to have strong innovative cultures, exhibited a significant deviation from the exponential distribution. For example, if selection and management of inventors were strong, we would expect to observe a concave up region in the upper tail of productivity. Only Phillips exhibited this concave up region and then only for the top 1% of inventors. None of the other companies exhibited such evidence of extra investment in high productivity inventors.

Furthermore, the range of the parameter of the exponential distribution was only 2:1. While the differences between the parameters are statistically significant, they may be substantially affected by other factors besides the effectiveness of managing inventive productivity.

We presented evidence that quality is related to quantity for inventors. Eminent inventors have productivity (patents per year) that is more than twice that for typical companies. Furthermore, the inventors in the top 10% of productivity produce over 50% of the patents for these companies. If high productivity were merely a matter of churning out low quality patents, then the management of these companies is making poor investments of the $7,000-$140,000 cost of filing and maintaining a patent.

Also surprising is that the longevity of inventors follows the exponential distribution. That is, most inventors have a very short inventive career. Short careers may reflect late starts as much as early terminations.

Using a new method for measuring randomness, we find that every company had a large majority of inventors with no significant trend in their patent production. Also, each company's inventors' time pattern of production generally followed the common Poisson distribution. The proportion of non-random inventors with a discernable trend was only 1.3 % on average and the largest was 2.5%. The largest trend overall was 3%, but many companies exhibited no trend at all. Thus, the vast majority of inventors produce patents at a constant rate and follow a standard statistical distribution for fluctuation over time. For these companies, there was no significant evidence of learning or decline.

The most substantial deviations occurred for those companies with known dramatic changes in business strategy, with a likely related change in technology, that impose slumps in patent production, and that are observed as fluctuating inventive productivity for some or all of the inventors. But downsizing alone does not necessarily result in large fluctuations in productivity, as shown in the goodness-of-fit to the Poisson distribution in Table C.5. Phillips downsized but had small deviation, whereas Unisys had dramatic changes in strategy in addition to downsizing and had large deviation.

The over-riding similarities among these companies are indeed surprising. But, we show that they are expected findings from the statistics of exceedances. That is, we propose that inventors are characterized by two factors, talent and tenacity. Talent is manifest by productivity. Tenacity is manifest by longevity. The statistics of exceedances predicts that the upper tails of these distributions will follow the Generalized Pareto distribution, of which the exponential distribution is an important special case. Thus, there is a theoretical basis for the findings of this study.

These specific results can be summarized into a simple rule: Talent is more important than management. While this rule is widely believed, this study is the first to support it with a systematic, analysis of objective facts. This rule is also supported by other studies that find that intrinsic motivation is more important to R&D workers than external motivations.[143] What is surprising is the small impact of management. These findings also support the allegation that society mostly "harvests creativity from the wild."[144]

The most significant implication for management is the small evidence for additional investment of resources (e.g. technicians, equipment, patent assistance, etc.) in high productivity inventors. If such investments were made, the upper tail of the distribution of productivity should be concave up. When it is not, it may be evidence that high productivity inventors are not receiving their appropriate share of resources. Thus, resources should not be distributed equally among all contenders. When that happens, the lowest productivity inventors are getting as much support as the highest productivity inventors. An argument against this investment strategy is that it favors the inventor who may just be lucky in the beginning. However, the talent level will ultimately prevail. Resources alone do not make inventions; it takes creative ideas first. We propose that improving inventive productivity also requires insightful and courageous management—men and women who will make investments according to inventive productivity. It takes this kind of manager to provide twice the technicians to the top 10% of inventors than are provided to the average inventor. Furthermore, it takes this kind of manager to advise the low productivity inventors to consider other career paths.

We do not want to appear to be critical of R&D managers. Indeed, the author has over 25 years experience in managing new product development. It is a difficult task and there have been few effective ways to measure performance. The methods described in this study may be a new aid to R&D management.

# C.11 Statistical Details

We have deliberately deferred most of the statistical details to this section because we believe that many readers will be more interested in the overall findings. A more detailed description is in the first endnote to this Appendix. The sub-headings in the abstract follow the order of statistical analysis and do not follow the order of the major sections in the body of this paper.

## C.11.1 Inventive Productivity

The minimum criteria of five patents over at least three years was chosen after extensive analysis of real data. When these thresholds are smaller, the distributions of inventive productivity are only weakly exponential, though they still follow the Generalized Pareto Distribution (GPD). However, the analysis of the GPD is much more complicated, so we accepted the values that make the analysis straightforward. Larger thresholds can be used, but the number of inventors begins to decline. The resulting smaller samples cause the statistical conclusions to be less powerful.

In each case, standard statistical methods were followed by confirming the exponential model with QQ plots[145] before proceeding with goodness of fit. For each sample, the QQ plot exhibited a substantial linear region above a small threshold, confirming the exponential model.

It is important to record the patents dates as accurately as possible. Patents are issued weekly and so the time precision is about 0.02 years. Thus, the time between the first and last observed patents has the same precision. This method allows inventive productivity to be a continuous variable and the statistical methods are more powerful. Conversely, if time is measured only in whole years, then inventive productivity is the ratio of two integers and is not a continuous variable and the statistical methods are weaker.

The goodness-of-fit is determined by the Cramer-von Mises test because its critical values are known when the exponential parameter is estimated from the empirical data.[146] The more well known Chi-Square test could be used, but is less powerful.[147]

The comparison of inventive productivity uses the F-test for the difference between standard deviations.[148] The more well known test using the normal distribution is not appropriate because of the exponential distribution is skewed.

## C.11.2 Randomness and Poisson-ness

The test of randomness examines the patent counts in time elements. While these time elements could be whole years, we have found that using time elements of one-third-year makes better use of the short career of many inventors. Consider our minimum criteria of five patents spread over three years. If whole years are used, there are only three time elements to compare for randomness. If one-third-year time elements are used, there are nine time elements to compare.

The test of randomness sums the correlations of the patent counts in nearest neighbor time elements, next-nearest neighbors, etc. We found that the most powerful choice was using nearest neighbor, second, and third neighbors.[149] However, that method requires that the data be purely random. Unfortunately, the inventor data always begins and ends with a time element with at least one patent, and so is not purely random.

This problem was solved by computer simulation. That is, 3,000 "inventor's" inventive productivities were drawn from an exponential distribution with the company's value of phi (e.g. 0.55 patents per year). Thus, each "inventor" has a specified inventive productivity (e.g. 0.333 patents per year). Then the patent counts for 100 time elements were drawn from the Poisson distribution with this productivity (e.g. 0.111 for patents per one-third-year). Then each "inventor" was assigned a career duration drawn from an exponential distribution with the company's value of tau (e.g. 13 years). These "inventors" were assigned a career start date such that the number of "inventors" was constant throughout the 1976-1999 time interval. Only those "inventors" with at least five patents over at least nine time elements were retained. Then any leading or trailing zeroes were eliminated, so that the simulated "inventors" have observed careers that start and end with a patent, matching the real inventors. Usually, about 1,000 "inventors" were then analyzed for randomness and Poisson-ness to produce the simulation curves in Figures C.4-C.7. Although each company's distribution of randomness and Poisson-ness differed from the simulations, the differences are generally small, as seen in Table C.5 and its related discussion.

Fluctuating inventive productivity was simulated by making each simulated "inventor's" inventive productivity fluctuate in a random manner over the 100 time elements. The simplest method is the Negative Binomial Distribution.[150] Each "inventor" retained the same average inventive productivity (e.g. 0.333), but the value for each time element was drawn from a gamma distribution. The gamma parameter, alpha, determined the degree of fluctuation. The simulation for Unisys in Figure C.7 has alpha value of three. Admittedly, we have no independent information that justifies using the Negative Binomial Distribution. It is just the simplest model for a fluctuating inventive productivity. However, the similarity of shape between the empirical Unisys distribution and the simulation shown in Figure C.7 supports this assumption.

## C.11.3 Inventive Longevity

The key issue in determining an inventor's longevity is estimating the time before the first patent and after the last patent. But, it is not so simple as taking the reciprocal of inventive productivity (e.g. 1/0.333 = 3 years) for each time estimate.

In the section above, we established that the Poisson distribution describes the time pattern of production for most inventors in most companies. The Poisson distribution has the important characteristic that the time-between-patents follows the exponential distribution[151] from which the Poisson parameter can be estimated.[152] (This relationship has nothing to do with the exponential distribution of inventive productivity or longevity.) The exponential distribution has the important characteristic that the mean is equal to the standard deviation. That is, the mean and standard deviation of the time-between-patents is equal to the reciprocal of an inventor's productivity (e.g. 1/0.333 = 3.0 years). But, the exponential distribution is skewed, as we showed in Figure C2. So the mean is not a good estimator of the typical event.

We have found that choosing the 0.80 probability of the time-between-patents is a useful criterion.[153] That is, we will consider that an inventor's career is completely recorded within the 1976-1999 time interval if there is at least 0.80 probability that another patent is not just beyond the time interval limits. An example will serve to clarify the situation. Consider an inventor with productivity of 0.53 patents per year (or 1/0.53 = 1.88 years-between-patents), first observed patent at Jan. 1, 1980 and last observed patent at Jan. 1, 1995. The exponential distribution 0.80 point is at 1.6x1.88 = 3.0 years. Thus, there is 0.80 probability that the first observed patent is truly that inventor's first patent if no patent occurs between Jan. 1, 1980 and Jan. 1, 1983. Similarly, there is 0.80 probability that the last observed patent is truly that inventor's last patent if no patent occurred between Jan. 1, 1995 and Jan. 1, 1998. Since both of these are true, we conclude that this inventor's entire career has been recorded. Furthermore, we estimate this inventor's longevity as 18 years (Jan. 1, 1980 to Jan. 1, 1998).

In each case, standard statistical methods were followed by confirming the exponential model with QQ plots[154] before proceeding with goodness of fit. For each sample, the QQ plot exhibited a substantial linear region above a small threshold, confirming the exponential model. Goodness-of-fit was determined by the Cramer-von Mises method for censored data.[155]

The statistics of exceedances describe the upper tails of common distributions. This is the justification for ignoring the inventors below the thresholds in productivity and longevity. That is, the threshold is the value necessary to identify the upper tail that corresponds to the methods for the statistics of exceedances. It is important to emphasize that only a minority of inventors (with at least five patents over three years) fell below these thresholds for each sample: about 28% for productivity and about 18% for longevity.

# *Index*

# *Notes*

## Preface

[1], Greg A., and Burley, James. 1997. 3,000 raw ideas = 1 commercial success! *Research-Technology Management* 40:16–27.

## Chapter 1. Introduction: Managing Innovation

[2] For more ways to examine the innovativeness of projects, read the following:

Unsworth, Kerrie. 2001. Unpacking creativity. *Academy of Management Review* 26:289–297.

Perkins, David N. 1992. The topography of invention. In Robert J. Weber and David N. Perkins (Eds.), *Inventive Minds: Creativity in Technology*, (pp. 238–250). New York: Oxford University Press

Austin, James H. 1978. *Chase, chance and creativity: The lucky art of novelty.* New York: Columbia University Press, pp. 72–77.

Pearson, Alan W. 1991. Managing innovation: An uncertainty reduction process. In Jane Henry and David Walker (Eds.) *Managing Innovation*, (pp. 18–27). Thousand Oaks, CA: Sage Publications

[3] For more information on breakthrough innovations, read this book.

Leifer, Richard, McDermott, Christopher M., O'Connor, Gina Colarelli, Peters, Lois S., Rice, Mark P., and Veryzer, Robert W. 2000. *Radical innovation: How mature companies can outsmart upstarts.* New York: Harvard Business School Press.

[4] For more information on incremental improvements versus innovations, read this book and these articles.

Porter, Michael F. 1980. *Competitive Strategy: Techniques for Analyzing Industries and Competitors.* New York: The Free Press, pp. 34–41.

Cooper, Robert G., Edgett, Scott J. and Kleinschmidt, Elko J. 1997. Portfolio management in new product development: Lessons from the leaders-II. *Research-Technology Management* 40:43–52.

Song, X. Michael, and Montoya-Weiss, Mitzi M. 1998. Critical development activities for really new versus incremental products. *Journal of Product Innovation Management* 15:124–135.

[5] For those readers who are interested in exploring the difference between risk and uncertainty, we recommend this book.

Bernstein, Peter L. 1996. *Against the gods: The remarkable story of risk.* New York: John Wiley and Sons.

[6] Though the importance of Uncommon Knowledge is frequently overlooked, it is not new.

Gilman, John J. 1992. *Inventivity: The art and science of research management.* New York: Van Nostrand Reinhold, p. 2.

Krogh, Lester C. 1987. Measuring and improving laboratory productivity/quality. *Research-Technology Management* 30:22–24.

Kao, John. 1996. *Jamming: The art and discipline of business creativity.* New York: HarperCollins.

[7] Here are some good readings on tacit knowledge.

Polanyi, Michael. 1983. *The tacit dimension.* New York: Peter Smith.

Reber, Arthur S. 1993. *Implicit learning and tacit knowledge: An essay on the cognitive unconscious.* New York: Oxford University Press.

Mascitelli, Ronald. 2000. From experience: Harnessing tacit knowledge to achieve breakthrough innovation. *Journal of Product Innovation Management* 17:179–193.

Wagner, Richard K., and Sternberg, Robert J. 1985. Practical intelligence in real-world pursuits: The role of tacit knowledge. *Journal of Personality and Social Psychology* 49:436–458.

Nonaka, Ikujiro, and Takeuchi, Hirotaka. 1995. *The knowledge-creating company: How Japanese companies create the dynamics of innovation.* New York: Oxford University Press.

Schon, Donald A. 1983. *The reflective pratitioner: How professionals think in action.* New York: Basic Books.

[8] I am indebted to Dr. Les Krogh, former vice president of R&D at 3M Company, for suggesting this topic.

[9] Here are just a few readings on the value of patents.

Narin, Francis, Noma, Elliot, and Perry, Ross. 1987. Patents as indicators of corporate technological strength. *Research Policy* 16:143–155.

Pegels, C. Carl, and Thirumurthy, M. V. 1996. The impact of technology strategy on firm performance. *IEEE Transactions on Engineering Management* 43:246–249.

Kortum, Samuel and Lerner, Josh. 1999. What is behind the recent surge in patenting? *Research Policy* 28:1–22.

Harhott, Dietmar, Narin, Francis, Scherer, F. M., and Katrin Vopel. 1999. Citation frequency and the value of patented inventions. *The Review of Economics and Statistics* 81:511–515.

Krogh, Lester C. 1987. Measuring and improving laboratory productivity/quality. *Research-Technology Management* 30:22–24.

Gilman, John J. 1992. *Inventivity: The art and science of research management.* New York: Van Nostrand Reinhold, p. 49.

Rivette, Kevin G. and Kline, David. 2000. *Rembrandts in the attic: Unlocking the hidden value of patents.* New York: Harvard Business School Press.

## Chapter 2. Mining for Important Market Problems

[10] Betz, Frederick. 1998. *Managing technological innovation: Competitive advantage from change.* New York: John Wiley and Sons, p. 116.

## Chapter 3. Mining for Inventive Ideas

[11] von Hippel, Eric. 1988. *The sources of innovation.* New York: Oxford University Press.

von Hippel, Eric. 1997. Product and process concept development via the lead user method. In Ralph Katz (Ed.), *The Human Side of Managing Technological Innovation,* (pp. 561–571). New York: Oxford University Press.

[12] Davidow, William H. 1986. *Marketing high technology: An insider's view.* New York: The Free Press., p. 37

[13] Davidow, William H. 1986. *Marketing high technology: An insider's view.* New York: The Free Press, p. 156.

[14] Huber, John C. 1999. Inventive productivity and the statistics of exceedances. *Scientometrics* 45:33–53.

[15] Huber, John C. in-press. Identifying prolific inventors: Results from the inventor profile. *Research-Technology Management*

[16] Ibid.

[17] Ibid.

[18] Lotka, A. J. 1926. The frequency of distribution of scientific productivity. *Journal of the Washington Academy of Science* 16:317–323.

Pao, Miranda Lee. 1986. An empirical examination of Lotka's law. *Journal of the American Society for Information Science* 37:26–33.

Huber, John C. in-press. A new method for analyzing scientific productivity. *Journal of the American Society for Information Science and Technology.*

Price, Derek J. De Solla. 1963. *Little science, big science.* New York: Columbia University Press.

[19] Since we only measure the rate of patent production for those 9% of inventors with at least five patents, this exponential relationship is really the upper tail of some other distribution, see Huber. It does fit the upper tail of the normal distribution. So we might say that inven-

tive talent has a normal distribution, similar to IQ, height, weight, etc. See Ibid, Huber.

[20] Huber, John C. 1999. Inventive productivity and the statistics of exceedances. *Scientometrics* 45:33–53.

[21] www.bpmlegal.com

[22] Coyne, William E., former Executive Vice President of Research and Development, 3M Company. Email to author, November 2, 2000.

[23] Barrett, Deidre. 2001. *The committee of sleep: How artists, scientists, and athletes use dreams for creative problem-solving—and how you can too.* New York: Crown Publishers.

Root-Bernstein, Robert Scott. 1989. *Discovering.* New York: Harvard University Press, Cambridge MA.

Root-Bernstein, Robert, & Root-Bernstein, Michele. 1999. *Sparks of genius.* New York: Houghton-Mifflin.

Shore, Sidney X. 1999. *Invent! Constructive discontent.* New York: Crisp Publications.

Siler, Todd. 1996. *Think like a genius.* New York: Bantam Books.

von Oech, Roger. 1986. *A kick in the seat of the pants.* New York: Harper & Row.

von Oech, Roger. 1983. *A whack on the side of the head.* New York: Warner Books.

[24] Chopra, A. J. 1999. *Managing the people side of innovation.* New York: Kumarian Press.

Blohowiak, Donald W. 1992. *Mavericks! How to lead your staff to think like Einstein, create like da Vinci, and invent like Edison.* New York: Business One Irwin.

Bramson, Robert M. 1981. *Coping with difficult people.* New York: Anchor Press/Doubleday.

[25] Ibid.

[26] The importance of rate of patent production (rather than total patents) is discussed in these readings.

Porter, Michael E., and Stern, Scott. 1999. *The new challenge to America's prosperity: Findings from the innovation index.* New York: Council on Competitiveness.

Scott, Susanne G., and Bruce, Reginald A. 1994. Determinants of innovative behavior: A path model of individual innovation in the workplace. *Academy of Management Journal* 37:580–607.

Huber, John C. 1999. Inventive productivity and the statistics of exceedances. *Scientometrics* 45:33–53.

[27] Private communication from Henry Petroski, author of nine books on the practice of engineering.

[28] May, Rollo. 1975. *The courage to create.* New York: WW Norton.

[29] Abra, Jock. 1997. *The motives for creative work: An inquiry with speculations about sports and religion.* New York: Hampton Press, p. 44.

[30] Much of the problem in studying general creativity is the difficulty of measuring it. The Inventor Profile is the first study that uses an objective, quantitative measure and studies one a specific kind of creativity.

Plucker, Jonathan A., & Runco, Mark A. 1998. The death of creativity measurement has been greatly exaggerated: Current issues, recent advances, and future directions in creativity assessment. *Roeper Review* 21:36–39.

## Chapter 4. Mining for Good Strategic Statements

[31] Schmitt, Roland W. 1991. The strategic measure of R&D. *Research-Technology Management* 34:13–16.

Szakonyi, Robert. 1994. Measuring R&D effectiveness—I. *Research-Technology Management* 37:27–32.

## Chapter 5. Mining for Good Projects

[32] Stevens, Greg A., and Burley, James. 1997. 3,000 raw ideas = 1 commercial success! *Research-Technology Management* 40:16–27.

Rosenfeld, Robert, and Servo, Jenny C. 1990. Facilitating innovation in large organizations. In Michael A. West and James L. Farr (Eds.), *Innovation and Creativity at Work: Psychological and Organizational Strategies* (pp. 251–264). New York: John Wiley

Glass, Doug. 1996. Hallmark cares enough to treat employees best. *Austin American-Statesman*, (June 9) p. F-2.

[33] Gee, Edwin A., and Tyler, Chaplin. 1976. *Managing innovation.* New York: John Wiley, p. 72.

[34] Few books on managing innovation discuss the Fuzzy Front End in depth. These are the best until now.

Reinertsen, Donald G. 1997. *Managing the design factory: A product developer's toolkit.* New York: The Free Press, pp. 215–216.

Cooper, Robert G., Edgett, Scott J., and Kleinschmidt, Elko J. 1998. *Portfolio management for new products.* New York: Perseus Books, pp. 16–17.

[35] The notion of reducing uncertainty is closely tied to increasing the quality of information, as discussed in these readings.

Pearson, Alan W. 1991. Managing innovation: An uncertainty reduction process. In Jane Henry and David Walker (Eds.) *Managing Innovation*, (pp. 18–27). Thousand Oaks, CA: Sage Publications

Gilman, John J. 1992. *Inventivity: The art and science of research management.* New York: Van Nostrand Reinhold, p. 2.

Patterson, Marvin L. with Sam Lightman. 1993. *Accelerating innovation: Improving the process of product development.* New York: Van Nostrand Reinhold, pp. 77–97.

Reinertsen, Donald G. 1997. *Managing the design factory: A product developer's toolkit.* New York: The Free Press, pp. 68–83.

Moenaert, Rudy K., and Souder, William E. 1990. An information transfer model for integrating marketing and R&D personnel in new product development projects. *Journal of Product Innovation Management* 7:91–107.

Rycroft, Robert W., and Kash, Don E. 2000. Steering complex innovation. *Research-Technology Management* 43:18–23.

[36] Cooper, Robert G. 1996. Overhauling the new product process. *Industrial Marketing Management* 25:465–482.

[37] Cooper, Robert G., Edgett, Scott J. and Kleinschmidt, Elko J. 1997. Portfolio management in new product development: Lessons from the leaders—II. *Research-Technology Management* 40:43–52.

Koen, Peter, Ajamian, Greg, Burkart, Robert, Clamen, Allen, Davidson, Jeffrey, D'Amore, Robb, Elkins, Claudia, Herald, Kathy, Incorvia, Michael, Johnson, Albert, Karol, Robin, Seibert, Rebecca, Slavejkov, Aleksandar, and Wagner, Klaus. 2001. Providing clarity and a common language to the "Fuzzy Front End." *Research-Technology Management* 44:46–55.

[38] Howard, William G. Jr., and Guile, Bruce R. (Eds.). 1992. *Profiting from innovation.* New York: The Free Press, p. 47.

[39] Foster, Richard N. 1986. *Innovation: The attacker's advantage.* New York: Summit Books.

[40] Tritle, Gary L., Scriven, Eric F. V., and Fusfeld, Alan R. 2000. Resolving uncertainty in R&D portfolios. *Research-Technology Management* 43:47–55.

[41] In my opinion, the best project selection system is described in Cooper, etal. The other readings also describe good systems.

Cooper, Robert G., Edgett, Scott J., and Kleinschmidt, Elko J. 1998. *Portfolio management for new products.* New York: Perseus Books, pp. 36–44.

Crawford, C. Merle. 1983. *New products management.* New York: Richard D. Irwin, pp. 366–385.

Hughes, G. David, and Chafin, Don C. 1996. Turning new product development into a continuous learning process. *Journal of Product Innovation Management* 13:89–104.

Yap, Chee Meng, and Souder, Wm. E. 1993. A filter system for technology evaluation and selection. *Technovation* 13:449–469.

[42] Udell, Lawrence J. 2001. Ask all the right questions. *Research-Technology Management* 44:13–14.

Cooper, Robert G. 1993. *Winning at new products (2nd ed).* New York: Perseus Books, p. 201.

[43] Tritle, Gary L., Scriven, Eric F. V., and Fusfeld, Alan R. 2000. Resolving uncertainty in R&D portfolios. *Research-Technology Management* 43:47–55.

Schrage, Michael., 2000., *Serious play: How the world's best companies simulate to innovate.*, New York: Harvard Business School Press, pp. 149–150.

[44] Wheelwright, Steven C., and Clark, Kim B. 1995. *Leading product development: The senior manager's guide to creating and shaping the enterprise.* New York: The Free Press, pp. 125–127.

[45] Chopra, A. J. 1999. *Managing the people side of innovation.* New York: Kumarian Press.

Blohowiak, Donald W. 1992. *Mavericks! How to lead your staff to think like Einstein, create like da Vinci, and invent like Edison.* New York: Business One Irwin.

[46] Peters, Thomas J., and Waterman, Robert H. Jr. 1982. *In search of excellence.* New York: Harper and Row, p. 229.

[47] Ibid, p. 203

[48] Reinertsen, Donald G. 1997. *Managing the design factory: A product developer's toolkit.* New York: The Free Press, pp. 11–14.

[49] Stevens, Greg A., and Burley, James. 1997. 3,000 raw ideas = 1 commercial success! *Research-Technology Management* 40:16–27.

Glass, Doug. 1996. Hallmark cares enough to treat employees best. *Austin American-Statesman*, (June 9): F-2.

Rosenfeld, Robert, and Servo, Jenny C. 1990. Facilitating innovation in large organizations. In Michael A. West and James L. Farr (Eds.), *Innovation and Creativity at Work: Psychological and Organizational Strategies*, (pp. 251–264). New York: John Wiley.

## Chapter 6. Mining for Successful New Products

[50] Gilman, John J. 1992. *Inventivity: The art and science of research management.* New York: Van Nostrand Reinhold, p. 2.

Patterson, Marvin L. with Sam Lightman. 1993. *Accelerating innovation: Improving the process of product development.* New York: Van Nostrand Reinhold, pp. 77–97.

Reinertsen, Donald G. 1997. *Managing the design factory: A product developer's toolkit.* New York: The Free Press, pp. 68–83.

Moenaert, Rudy K., and Souder, William E. 1990. An information transfer model for integrating marketing and R&D personnel in new product development projects. *Journal of Product Innovation Management* 7:91–107.

Rycroft, Robert W., and Kash, Don E. 2000. Steering complex innovation. *Research-Technology Management* 43:18–23.

[51] Cooper, Robert G. 1993. *Winning at new products (2nd ed).* New York: Perseus Books , pp. 106–120.

Wheelwright, Steven C., and Clark, Kim B. 1995. *Leading product development: The senior manager's guide to creating and shaping the enterprise.* New York: The Free Press, pp. 133–164.

[52] Cooper, Robert G. 1999. The invisible success factors in product innovation. *Journal of Product Innovation Management* 16:115-133.

[53] Cooper, Robert G. 1996. Overhauling the new product process. *Industrial Marketing Management* 25:465–482.

[54] Kay, Ronald. 1990. *Managing creativity in science and hi-tech.* New York: Springer-Verlag, pp. 30–32.

[55] White, Daniel. 1996. Stimulating innovative thinking. *Research-Technology Management* 39:31–35.

[56] Reinertsen, Donald G. 1997. *Managing the design factory: A product developer's toolkit.* New York: The Free Press, p.71.

[57] A more refined analogy to the topology of the gold fields is described in the following:

Perkins, David N. 1992. The topography of invention. In Robert J. Weber and David N. Perkins (Eds.), *Inventive Minds: Creativity in Technology*, (pp. 238–250). New York: Oxford University Press.

Perkins, David N. 1998. In the country of the blind: An appreciation of Donald Campbell's vision of creative thought. *Journal of Creative Behavior* 32:177–191.

[58] Reinertsen, Donald G. 1997. *Managing the design factory: A product developer's toolkit.* New York: The Free Press, p.71.

[59] Coyne, William E. 2001. How 3M innovates for long-term growth. *Research-Technology Management* 44:21–24.

[60] For more information about the conflicts between R&D and marketing, read the following:

Griffin, Abbie, and Hauser, John R. 1996. Integrating R&D and marketing: A review and analysis of the literature. *Journal of Product Innovation Management* 13:191–215.

Song, X. Michael, Neeley, Sabrina M., and Zhao, Yuzhen. 1996. Managing R&D - marketing integration in the new product development process. *Industrial Marketing Management* 25:545–553.

Griffin, Abbie, and Hauser, John R. 1992. Patterns of communication among marketing, engineering and manufacturing—a comparison between two product teams. *Management Science* 38:360–73.

Dougherty, Deborah. 1992. Interpretive barriers to successful product innovation in large firms. *Organization Science* 3:179–202.

Gupta, Ashok K., and Wilemon, David. 1988. The credibility-cooperation connection at the R&D-marketing interface. *Journal of Product Innovation Management* 5:20–31.

Gupta, Ashok K., and Wilemon, David. 1988. Why R&D resists using marketing information. *Research Management* 17:36–41.

Souder, William E. 1988. Managing relations between R&D and marketing in new product development projects. *Journal of Product Innovation Management* 5:6–19.

Gupta, Ashok K., Raj, S. P., and Wilemon, David. 1987. Managing the R&D-marketing interfac. *Research Management* 16:38–43.

Gupta, Ashok K., Raj, S. P., and Wilemon, David. 1985. The R&D-marketing interface in high-technology firms. *Journal of Product Innovation Management* 2:12–24.

[61] Kay, Ronald. 1990. *Managing creativity in science and hi-tech.* New York: Springer-Verlag, pp. 48–49.

[62] Schrage, Michael. 2000. *Serious play: How the world's best companies simulate to innovate.* New York: Harvard Business School Press.

Reinertsen, Donald G. 1997. *Managing the design factory: A product developer's toolkit.* New York: The Free Press.

[63] Edelheit, Lewis S. 1997. General Electric. In Rosabeth Moss Kanter, John Kao, and Fred Wiersema (Eds.), *Innovation: Breakthrough Thinking at 3M, DuPont, GE, Pfizer, and Rubbermaid*, (pp. 99–121). New York: HarperCollins, p. 10.

Stevens, Greg A., and Burley, James. 1997. 3,000 raw ideas = 1 commercial success! *Research-Technology Management* 40:16–27.

[64] Reinertsen, Donald G. 1997. *Managing the design factory: A product developer's toolkit.* New York: The Free Press, p. 174

[65] Cooper, Robert G. 1993. *Winning at new products (2nd ed).* New York: Perseus Books, pp. 150–153.

Wheelwright, Steven C., and Clark, Kim B. 1995. *Leading product development: The senior manager's guide to creating and shaping the enterprise.* New York: The Free Press, pp. 229–234.

Nuese, Charles J. 1995. *Building the right things right.* New York: The Kraus Organization.

Endres, Al. 1997. *Improving R&D performance the Juran way.* New York: John Wiley and Sons.

[66] Wheelwright, Steven C., and Clark, Kim B. 1995. *Leading product development: The senior manager's guide to creating and shaping the enterprise.* New York: The Free Press, pp. 105–110.

[67] Cooper, Robert G. 1993. *Winning at new products (2nd ed)*. New York: Perseus Books, pp. 333–334.

[68] Hall, John A. 1991. *Bringing new products to market: The art and science of creating winners*. New York: American Management Assn., p. 229.

Davidow, William H. 1986. *Marketing high technology: An insider's view*. New York: The Free Press.

[69] Often the differences are even larger.

Coyne, William E. 2001. How 3M innovates for long-term growth. *Research-Technology Management* 44:21–24.

Reinertsen, Donald G. 1997. *Managing the design factory: A product developer's toolkit*. New York: The Free Press, p. 39.

[70] Stevens, Greg A., and Burley, James. 1997. 3,000 raw ideas = 1 commercial success! *Research-Technology Management* 40:16–27.

## Chapter 7. Mining for Improved Future Innovations

[71] Coyne, William E. 2001. How 3M innovates for long-term growth. *Research-Technology Management* 44:21–24.

[72] Below are several good resources for a lessons-learned database.

Lynn, Gary S., Mazzuca, Mario, Morone, Joseph G., and Paulson, Albert S. 1998. Learning is the critical success factor in developing truly new products. *Research-Technology Management* 41:45–51.

Perry, Tekla S. 1995. How small firms innovate: Designing a culture for creativity. *Research-Technology Management* 38:14–7.

Reinertsen, Donald G. 1997. *Managing the design factory: A product developer's toolkit*. New York: The Free Press, pp. 141–143.

Wheelwright, Steven C., and Clark, Kim B. 1995. *Leading product development: The senior manager's guide to creating and shaping the enterprise*. New York: The Free Press, pp. 285–310.

[73] N. L. Johnson, S. Kotz, and N. Balakrishnan, *Continuous Univariate Distributions* (2nd ed 1). New York: John Wiley and Sons, 1994, p. 508.

M. Engelhardt, Reliability estimation and applications, in *The Exponential Distribution*, N. Balakrishnan and A. B. Basu, Eds. New York: Gordon and Breach, 1995:75.

[74] Gilman, John J. 1992. *Inventivity: The art and science of research management.* New York: Van Nostrand Reinhold, pp. 139–148.

[75] Although effective training has not been proved, at least at the level of a 20% improvement in productivity, the content of the following programs is promising.

Couger, J. Daniel. 1995. *Creative problem solving and opportunity finding.* New York: Boyd and Fraser.

Ford, Cameron M., and Gioia, Dennis A. (Eds.). 1995. *Creative action in organizations: Ivory tower visions and real world voices.* New York: Sage Publications, pp. 350–365.

VanGundy, Arthur B. 1992. *Idea power: Techniques and resources to unleash the creativity in your organization.* New York: American Management Assn.

VanGundy, Arthur B. 1987. *Creative problem solving: A guide for trainers and management.* New York: Quorum Books.

Kivenson, Gilbert. 1982. *The art and science of inventing.* New York: Van Nostrand Reinhold.

Parnes, Sidney J., Noller, Ruth B., and Biondi, Angelo M. 1977. *Guide to creative action.* New York: Charles Scribner's Sons.

Parnes, Sidney J. 1967. *Programming creative behavior.* New York: State University of New York Press.

[76] Rossman, Joseph. 1931. *The psychology of the inventor: A study of the patentee.* New York: The Inventor's Publishing Co., p. 217.

Parmerter, S. M., and Garber, J. D. 1971. Creative scientists rate creativity factors. *Research Management* 14:65–70.

[77] Perkins, D. N. 1981. *The mind's best work*. Cambridge MA: Harvard University Press, p. 273.

Perkins, David N. 1990. The nature and nurture of creativity. In Beau Fly Jones and Lorna Idol (Eds.), *Dimensions of Thinking and Cognitive Instruction*, (pp. 380–431).Hillsdale, NJ:Lawrence Erlbaum.

[78] Polya, G. 1957. *How to solve it: A new aspect of mathematical method, 2nd ed*. New York: Princeton University Press.

Polya, G. 1954. *Mathematics and plausible reasoning: Vol. I induction and analogy in mathematics.* New York: Princeton University Press.

Polya, G. 1954. *Mathematics and plausible reasoning: Vol. II patterns of plausible inference.* New York: Princeton University Press.

Weber, Robert J., and Perkins, David N. (Eds.). 1992. *Inventive minds: Creativity in technology.* New York: Oxford University Press.

[79] Dasgupta, Subrata. 1994. *Creativity in invention and design: Computational and cognitive explorations of technological originality.* New York: Cambridge University Press, p. 26.

Sternberg, Robert J. 1996. *Successful intelligence: How practical and creative intelligence determine success in life.* New York: Simon and Schuster, pp. 200–218.

[80] Born to tinker. 2001. *3M Stemwinder* (January 23): 4

[81] MacLachlan, Alexander. 1995. Creativity in a large company: All you have to do is ask for it. In Cameron M. Ford and Dennis A. Gioia (Eds.), *Creative Action in Organizations: Ivory Tower Visions and Real World Voices*, (pp. 200–215). Thousand Oaks, CA:Sage Publications

Scott, Susanne G., and Bruce, Reginald A. 1994. Determinants of innovative behavior: A path model of individual innovation in the workplace. *Academy of Management Journal* 37:580–607.

Miller, Joseph. 1997. E. I. duPont deNemours and company, inc. In Rosabeth Moss Kanter, John Kao, and Fred Wiersema, *Innovation: Breakthrough Thinking at 3M, DuPont, GE, Pfizer, and Rubbermaid*, (pp. 65–71). New York: HarperCollins, p.70.

Coyne, William E. 2001. How 3M innovates for long-term growth. *Research-Technology Management* 44:21–24.
[82] Huber, John C. 1999. Inventive productivity and the statistics of exceedances. *Scientometrics* 45:33–53.
[83] Gilbert, Thomas F. 1978. *Human competence: Engineering worthy performance.* New York: McGraw-Hill.
[84] Huber, John C. 1999. Inventive productivity and the statistics of exceedances. *Scientometrics* 45:33–53.
[85] Mandino, Og. 1972. *The greatest secret in the world.* New York: Frederick Fell Publishing, p. 130.

## Chapter 8. Mining for Better Management of Innovation

[86] De Pree, Max. 1989. *Leadership is an art.* New York: Doubleday.
[87] Kouzes, James M., and Posner, Barry J. 1993. *Credibility: How leaders gain and lose it, why people demand it.* New York: Jossey-Bass.
[88] Bennis, Warren, and Goldsmith, Joan. 1994. *Learning to lead: A workbook on becoming a leader.* New York: Addison-Wesley.
[89] Brown, Mark G., and Svenson, Raynold A. 1998. Measuring R&D productivity. *Research-Technology Management* 41:30–35.

## Chapter 9. Mining for Companies That Manage Innovation Better

[90] Cooper, Robert G. 1999. The invisible success factors in product innovation. *Journal of Product Innovation Management* 16:115-133.
[91] Cooper, Robert G., and Kleinschmidt, Elko J. 1995. Benchmarking the firm's critical success factors in new product development. *Research-Technology Management* 38:374-391.
[92] Huber, John C. in-press. A new method for analyzing scientific productivity. *Journal of the American Society for Information Science & Technology*

## Appendix B. The Inventor Profile

[93] This Appendix was originally published as "Identifying Prolific Inventors: Results of the Inventor Profile" in *Research-Technology Management*, 2001.

[94] Birr, Kendall. 1957. *Pioneering in industrial research: The story of the General Electric research laboratory.* New York: The Public Affairs Press.

Broderick, John T. 1945. *Willis Rodney Whitney: Pioneer of industrial research.* New York: Fort Orange Press.

Kelley, Robert, and Caplan, Janet. 1993. How Bell Labs creates star performers. *Harvard Business Review*, (July–Aug):128–139.

Kanter, Rosabeth Moss, Kao, John, and Wiersema, Fred. 1997. *Innovation: Breakthrough thinking at 3M, duPont, GE, Pfizer, and Rubbermaid.* New York: HarperCollins.

Amabile, Teresa M., Conti, Regina, Coon, Heather, Lazenby, Jeffrey, and Herron, Michael. 1996. Assessing the work environment for creativity. *Academy of Management Journal* 39:1154–1184.

Cummings, Anne, and Oldham, Greg R. 1997. Enhancing creativity: Managing work contexts for the high potential employee. *California Management Review* 40:22–38.

[95] Busse, Thomas V., and Mansfield, Richard S. 1984. Selected personality traits and achievement in male scientists. *Journal of Psychology* 116:117–31.

Keller, Robert T., and Holland, Winford E. 1978. A cross-validation study of the Kirton adaption-innovation inventory in three research organizations. *Applied Psychological Measurement* 2:563–570.

Keller, Robert T., and Holland, Winford E. 1978. Individual characteristics of innovativeness and communication in research and development organizations. *Journal of Applied Psychology* 63:759–762.

Keller, Robert T., and Holland, Winford E. 1979. Toward a selection battery for research and development employees. *IEEE Transactions on Engineering Management* 26:90–93.

[96] 3M Staffing Resource Center. 1997. Hiring innovators. *Research-Technology Management* 40:8.

Johnson, Marvin M. 1996. Finding creativity in a technical organization. *Research-Technology Management* 39:9–11.

Roe, Anne. 1951. A psychological study of physical scientists. *Genetic Psychology Monographs* 43:121–239.

Dennard, Robert H. 2000. Creativity in the 2000s and beyond. *Research-Technology Management* 43:23–25.

Pawlak, Andrzej M. 2000. Fostering creativity in the new millennium. *Research-Technology Management* 43:32–35.

Pelz, Donald C., and Andrews, Frank M. 1976. *Scientists in organizations: Productive climates for research and development.* New York: Institute for Social Research, The University of Michigan.

[97] Huber, John C. 2000. A statistical analysis of special cases of creativity. *Journal of Creative Behavior* 34:203–225.

[98] Simonton, Dean Keith. 1988. *Scientific genius: A psychology of science.* New York: Cambridge University Press, p. 188.

[99] Cooper, Robert G., Edgett, Scott J., and Kleinschmidt, Elko J. 1998. *Portfolio management for new products.* New York: Perseus Books, p. 36–44.

[100] Huber, John C. 1999. Inventive productivity and the statistics of exceedances. *Scientometrics* 45:33–53.

Huber, John C. 2000. A statistical analysis of special cases of creativity. *Journal of Creative Behavior* 34:203–225.

[101] Ibid, 1999.

[102] Born to tinker. 2001. *3M Stemwinder* (January 23): 4

MacLachlan, Alexander. 1995. Creativity in a large company: All you have to do is ask for it. In Cameron M. Ford and Dennis A. Gioia (Eds.), *Creative Action in Organizations: Ivory Tower Visions and*

*Real World Voices*, (pp. 200–215). Thousand Oaks, CA:Sage Publications

[103] Owens, W. A. 1969. Cognitive, noncognitive, and environmental correlates of mechanical ingenuity. *Journal of Applied Psychology* 53:199–208.

[104] Petroski, Henry, author of nine books on the practice of engineering. Email to author, March 22, 2001.

[105] Mascitelli, Ronald. 2000. From experience: Harnessing tacit knowledge to achieve breakthrough innovation. *Journal of Product Innovation Management* 17:179–193.

Wagner, Richard K., and Sternberg, Robert J. 1985. Practical intelligence in real-world pursuits: The role of tacit knowledge. *Journal of Personality and Social Psychology* 49:436–458.

[106] Coyne, William E., former Executive Vice President of Research and Development, 3M Company. Email to author, November 2, 2000.

## Appendix C. Patent Productivity of Individuals and Companies

[107] Huber, J. C. 2001. A new method for analyzing scientific productivity. *Journal of the American Society for Information Science and Technology* 52:in-press.

[108] Shockley, W. 1957.On the statistics of individual variations of productivity in research laboratories. *Proceedings of the IRE* 47:279-290.

Albert, M. B., Avery, D., Narin, F., and McAllister, P. 1991. Direct validation of citation counts as indicators of industrially important patents, *Research Policy* 20:251-259.

Deng, Z., Lev, B., and Narin, F. 1999. Science and technology as predictors of stock performance. *Association for Investment Management and Research* 55:20-32, May-Jun.

Narin, F. and Breitzman, A. Inventive productivity, *Research Policy* 24:507-519, 1995.

F. Narin, M. P. Carpenter, and P. Woolf, 1984. Technological performance assessments based on patents and patent citations. *IEEE Transactions on Engineering Management* 31:172-183.

Huber, J. C. 1999. Inventive productivity and the statistics of exceedances, *Scientometrics.* 45:33-53.

Ernst, H., Leptien, C., and Vitt, J. 2000. Inventors are not alike: The distribution of patenting output among industrial R&D personnel. *IEEE Transactions on Engineering Management* 47:184-199.

Rossman, J. 1964. *Industrial Creativity: The psychology of the inventor.* New Hyde Park, NY: University Books.

[109] Roberts, E. B. 1988. What we've learned: Managing invention and innovation. *Research-Technology Management* 31:11-29, Jan-Feb.

Rossman, J. 1931. *The Psychology of the Inventor: A Study of the Patentee.* Washington, DC: The Inventor's Publishing Co.

Pelz, D. C. and Andrews, F. M. 1976. *Scientists in Organizations:Productive Climates for Research and Development.* Ann Arbor, MI: Institute for Social Research, The University of Michigan.

Keller, R. T. and Holland, W. E. 1978. Individual characteristics of innovativeness and communication in research and development organizations. *Journal of Applied Psychology* 63:759-762.

Keller, R. T. and Holland, W. E. 1979. Toward a selection battery for research and development employees. *IEEE Transactions on Engineering Management* 26:90-93.

Keller, R. T. and Holland, W. E. 1982. The measurement of performance among research and development professional employees: A longitudinal analysis. *IEEE Transactions on Engineering Management* 29:54-58.

Amabile, T. M. 1988. A model of creativity and innovation in organizations. *Research in Organizational Behavior* 10:123-167.

Amabile, T. M. 1997. Entrepreneurial creativity through motivational synergy. *Journal of Creative Behavior* 31:18-26.

Amabile, T. M. and Gryskiewicz, S. S. 1988. Creative human resources in the R&D laboratory: how environment and personality affect innovation. in *Handbook for Creative and Innovative Managers*, R. L. Kuhn, Ed. New York: McGraw-Hill, 501-524.

Simonton, D. K. 1992. The social context of career success and course for 2,026 scientists and inventors. *Personality and Social Psychology Bulletin* 18:452-63.

[110] Johnson, M. M. 1996. Finding creativity in a technical organization. *Research-Technology Management* 39:9-11, Sep-Oct.

Wolff, M. 1979. How to find—and keep—creative people. *Research Management* 22:43-45, Sep.

Krogh, L. C., Prager, J. H., Sorenson, D. P., and Tomlinson, J. D. 1988. How 3M evaluates its R&D programs. *Research-Technology Management* 31:10-14, Nov-Dec.

Kanter, R. M., Kao, J. and Wiersema, F. 1997. *Innovation: Breakthrough Thinking at 3M, DuPont, GE, Pfizer, and Rubbermaid.* New York: HarperCollins.

Schmitt, R. W. 1991. The strategic measure of R&D. *Research-Technology Management* 34:13-16, Nov-Dec.

[111] Roberts, E. B. 1988. What we've learned: Managing invention and innovation. *Research-Technology Management* 31:11-29, Jan-Feb.

Tornatzky, L. G. and Fleischer, M. 1990. *The Process of Technological Innovation.* Lexington, MA: Lexington Books.

Whiteley, R., Parish, T., Dressler, R. and Nicholson, G. 1998. Evaluating R&D performance using the new sales ratio. *Research-Technology Management* 41:20-22, Sep-Oct.

[112] Cooper, R. G. 1979. The dimensions of industrial new product success and failure. *Journal of Marketing* 43:93-103.

Cooper, R. G. and Kleinschmidt, E. J. 1993. Uncovering the keys to new product success. *IEEE Engineering Management Review*:5-18, Winter.

[113] Cooper, R. G. 1999. The invisible success factors in product innovation. *Journal of Product Innovation Management* 16:115-133.
[114] Souder, W. E. 1987. *Managing New Product Innovations.* Lexington, MA: Lexington Books, p. 243
Drucker, P. F. 1985. *Innovation and Entrepreneurship.* New York: Harper and Row.
von Hippel, E. 1988. *The Sources of Innovation.* New York: Oxford University Press.
von Hippel, E. 1997. Product and process concept development via the lead user method. in *The Human Side of Managing Technological Innovation,* Ralph Katz, Ed. New York: Oxford University Press, 561-571.
[115] Albert, M. B., Avery, D., Narin, F., and McAllister, P. 1991. Direct validation of citation counts as indicators of industrially important patents. *Research Policy* 20:251-259.
Deng, Z., Lev, B., and Narin, F. 1999. Science and technology as predictors of stock performance. *Association for Investment Management and Research* 55:20-32, May-Jun.
Pegels, C. C. and Thirumurthy, M. V. 1996. The impact of technology strategy on firm performance. *IEEE Transactions on Engineering Management* 43:246-249.
Porter, M. E. and Stern, S. 1999. *The New Challenge to America's Prosperity: Findings from the Innovation Index.* Washington, DC: Council on Competitiveness.
[116] Jolly, V. K. 1997. *Commercializing New Technologies: Getting from Mind to Market.* Boston, MA: Harvard Business School Press, p. 113.
Stevens, G. A. and Burley, J. 1997. 3,000 raw ideas = 1 commercial success! *Research-Technology Management* 40:16-27, May-Jun.
[117] Gowan, J. C. 1977. Some new thoughts on the development of creativity. *Journal of Creative Behavior* 11:77-90.

[118] Kanter, R. M., Kao, J., and Wiersema, F. 1997. *Innovation: Breakthrough Thinking at 3M, DuPont, GE, Pfizer, and Rubbermaid.* New York: HarperCollins.

Krogh, L. C., Prager, J. H., Sorenson, D. P., and Tomlinson, J. D. 1988. How 3M evaluates its R&D programs. *Research-Technology Management* 31:10-14, Nov-Dec.

Allio, M. K. 1993. 3M's sophisticated formula for teamwork, *Planning Review* 21:19-21, Nov-Dec.

Gundling, E. 2000. *The 3M Way to Innovation: Balancing People and Profit.* New York: Kodansha International.

Gupta, A. and Singhal, A. 1993. Managing human resources for innovation and creativity. *Research-Technology Management* 36:41-48, May-Jun.

Schmitt, R. W. 1991. The strategic measure of R&D. *Research-Technology Management* 34:13-16, Nov-Dec.

[119] Narin, F. and Breitzman, A. 1995. Inventive productivity. *Research Policy* 24:507-519.

[120] Drucker, P. F. 1946. *Concept of the Corporation.* New York: The John Day Co.

De Lorean, J. Z. 1979. *On a Clear Day You Can See General Motors.* Grosse Point, MI: Wright Enterprises.

Hirshberg, J. 1998. *The Creative Priority: Driving innovative business in the real world.* New York: Harper Business.

[121] Hooper, L. 1992. Unruh saves Unisys, now aims to put it on cutting edge. *Wall Street Journal*, p. B3, Sep 25.

Weber, J. 1989. This is hardly the turning point Unisys had in mind. *Business Week* 60:83-84, Aug 28.

Weinberg, N. 1995. Two Unisys execs resign over services strategy, *Computer World* 29:30, May 12.

[122] Birr, K. 1957. *Pioneering in Industrial Research: The Story of the General Electric Research Laboratory.* Washington, DC: The Public Affairs Press.

Broderick, J. T. 1945. *Willis Rodney Whitney: Pioneer of Industrial Research.* Albany NY: Fort Orange Press.

[123] O'Boyle, T. F. 1998. *At Any Cost: Jack Welch, General Electric, and the Pursuit of Profit.* New York: Random House.

Slater, R. 1999. *Jack Welch and the GE Way.* New York: McGraw-Hill.

[124] Edelheit, L. S. 1997. General Electric. in *Innovation: Breakthrough Thinking at 3M, DuPont, GE, Pfizer, and Rubbermaid*, R. M. Kanter, J. Kao, and F. Wiersema, Eds. New York: HarperCollins, 99-121.

[125] Johnson, M. M. 1996. Finding creativity in a technical organization. *Research-Technology Management* 39:9-11, Sep-Oct.

[126] Allison, D. 1987. When Phillips R&D had to cut back. *Research Management* 16:44-45, Mar-Apr.

[127] Porter, M. E. and Stern, S. 1999. *The New Challenge to America's Prosperity: Findings from the Innovation Index.* Washington, DC: Council on Competitiveness.

[128] Hewett, J. E. 1995. Two-stage and multi-stage tests of hypotheses. in *The Exponential Distribution,* N. Balakrishnan and A. B. Basu, Eds. New York: Gordon and Breach, 453-459.

[129] Zuckerman, H. 1977. *Scientific Elite: Nobel Laureates in the United States.* New York: The Free Press, p. 302.

[130] See the related reference in Appendix B.

[131] Simonton, D. K. 1997. Creative productivity: A predictive and explanatory model of career trajectories and landmarks. *Psychological Review* 104:66-89.

[132] Bain, L. J. and Engelhardt, M. 1991. *Statistical Analysis of Reliability and Life-testing Models: Theory and Methods.* New York: Marcel Dekker.

Elandt-Johnson, R. C. and Johnson, N. L. 1980. *Survival Models and Data Analysis.* New York: John Wiley and Sons.

Elsayed, E. A. 1996. *Reliability Engineering.* New York: Addison Wesley Longman.

Jewell, N. P., Kimber, A. C., Lee, M. T., and Whitmore, G. A. 1996. *Lifetime Data: Models in Reliability and Survival Analysis.* Boston, MA: Kluwer Academic.

Klein, J. P. and Moeschberger, M. L. 1997. *Survival Analysis: Techniques for Censored and Truncated Data.* New York: Springer-Verlag.

[133] Huber, J. C. 1999. Inventive productivity and the statistics of exceedances, *Scientometrics* 45:33-53.

[134] Allison, P. D., Long, J. S., and Krauze, T. K. 1982. Cumulative advantage and inequality in science. *American Sociological Review* 47:615-625.

Huber, J. C. 1998. Cumulative advantage and success-breeds-success: The value of time pattern analysis. *Journal of the American Society for Information Science* 49:471-476.

Price, D. S. 1976. A general theory of bibliometric and other cumulative advantage processes. *Journal of the American Society for Information Science* 27:292-306.

Glaenzel, W. and Schubert, A. 1990. The cumulative advantage function. A mathematical formulation based on conditional expectations and its application to Scientometric distribution. in *Informetrics 89/90*, L. Egghe and R. Rousseau, Eds. New York: Eslevier, 139-147.

[135] Huber, J. C. 2000. Portmanteau test for randomness in Poisson data. *Communications in Statistics: Simulation and Computation* 29:1165-1182.

[136] Boswell, M. T. 1986. Poisson index of dispersion. in *Encyclopedia of Statistical Sciences* 7, S. Kotz and N. L. Johnson, Eds. New York: John Wiley and Sons, 25-26.

Johnson, N. L., Kotz, S., and Kemp, A. W. 1993. *Univariate Discrete Distributions* (2nd ed). New York: John Wiley and Sons, p. 172.

[137] Rivette, K. G. and Kline, D. 2000. *Rembrandts in the Attic: Unlocking the Hidden Value of Patents.* Boston, MA: Harvard Business School Press, 106-110.

[138] Huber, J. C. 2000. A statistical analysis of special cases of creativity. *Journal of Creative Behavior* 34:203-225.

[139] Pickands, J. 1975. Statistical inference using extreme order statistics. *The Annals of Statistics* 3:119-131.

[140] Johnson, N. L., Kotz, S., and Balakrishnan, 1994. N. *Continuous Univariate Distributions* (2nd ed 1). New York: John Wiley and Sons, 614-620.

[141] Huber, J. C. 1999. Inventive productivity and the statistics of exceedances, *Scientometrics* 45:33-53.

[142] Huber, J. C. 2000. A statistical analysis of special cases of creativity, *Journal of Creative Behavior* 34:203-225.

Huber J. C. and Wagner-Dobler, R. 2001. Scientific production: A statistical analysis of authors in mathematical logic. *Scientometrics* 50:323-337.

Huber J. C. and Wagner-Dobler, R. 2001. Scientific production: A statistical analysis of authors in physics, 1800-1900. *Scientometrics* 50:437-453.

Huber, J. C. 2001. A new method for analyzing scientific productivity. *Journal of the American Society for Information Science and Technology* 51, in-press.

[143] Amabile, T. M. 1988. A model of creativity and innovation in organizations. *Research in Organizational Behavior* 10:123-167.

Amabile, T. M. 1997. Entrepreneurial creativity through motivational synergy. *Journal of Creative Behavior* 31:18-26.

[144] Gowan, J. C. 1977. Some new thoughts on the development of creativity. *Journal of Creative Behavior* 11:77-90.

[145] D'Agostino, R. B. 1986. Graphical analysis. in *Goodness of Fit Techniques,* R. B. D'Agostino and M. A. Stephens, Eds. New York: Marcel Dekker, 461-496.

[146] Stephens, M. A. 1986. Tests based on EDF statistics. in *Goodness of Fit Techniques,* R. B. D'Agostino and M. A. Stephens, Eds. New York: Marcel Dekker, 97-193.

[147] Moore, D. S. 1986. Tests of Chi-squared type. in *Goodness of Fit Techniques,* R. B. D'Agostino and M. A. Stephens, Eds. New York: Marcel Dekker, 63-96.

[148] Hewett, J. E. 1995. Two-stage and multi-stage tests of hypotheses. in *The Exponential Distribution*, N. Balakrishnan and A. B. Basu, Eds. New York: Gordon and Breach, 453-459.

[149] Huber, J. C. 2000. Portmanteau test for randomness in Poisson data, *Communications in Statistics: Simulation and Computation* 29:1165-1182.

[150] Johnson, N. L., Kotz, S., and Kemp, A. W. 1993. *Univariate Discrete Distributions* (2nd ed). New York: John Wiley and Sons, p. 204.

[151] Johnson, N. L., Kotz, S., and Kemp, A. W. 1993. *Univariate Discrete Distributions* (2nd ed). New York: John Wiley and Sons, p. 153.

[152] Johnson, N. L., Kotz, S., and Balakrishnan, N. 1994. *Continuous Univariate Distributions* (2nd ed 1). New York: John Wiley and Sons, p. 508.

Engelhardt, M. 1995. Reliability estimation and applications. in *The Exponential Distribution*, N. Balakrishnan and A. B. Basu, Eds. New York: Gordon and Breach, p. 75.

[153] Huber, J. C. 2000. A statistical analysis of special cases of creativity. *Journal of Creative Behavior* 34:203-225.

[154] D'Agostino, R. B. 1986. Graphical analysis. in *Goodness of Fit Techniques,* R. B. D'Agostino and M. A. Stephens, Eds. New York: Marcel Dekker, 461-496.

[155] Koziol, J. A. 1980. Goodness-of-fit tests for randomly censored data, *Biometrika* 67:693-696.